GW01605336

Warne's Transport Library

American Trucks of the Seventies

compiled by Elliott Kahn

edited by G. N. Georgano

FREDERICK WARNE

Published by Frederick Warne (Publishers) Ltd London 1981

Also in Warne's Transport Library

Farm Tractors by Nick Baldwin
English Horse Drawn Vehicles by David Parry
French Cars from 1920–25 by Pierre Dumont
Blue Blood by Serge Bellu
The GMC—a Universal Truck by J-M Boniface and J-G Jeudy
Trucks of the Sixties and Seventies by Nick Baldwin

ISBN 0 7232 2765 9

Filmset and printed in Great Britain by
BAS Printers, Limited, Over Wallop, Hampshire

CONTENTS

Abbreviations commonly used in the text

BBC = bumper to back of cab measurement
CID = cubic inch displacement, the usual American way of referring to engine capacity
GVW = gross vehicle weight, the weight of a truck plus its load. Sometimes GCW (gross combination weight) is used for tractor trailer trucks.
GPM = gallons per minute, referring to pumping capacity of fire pumps.
LPG = liquid propane gas

All prices are quoted in dollars; the exchange rate varied during the 1970s, starting at around **$**2·39 to the pound in 1970, dropping to a low of **$**1·60 in mid-1976, and rising to around **$**2·20 at the end of the decade.

PREFACE

Powered trucks in America, like much of the rest of the world, started with scattered experiments in the 1800s, but it was not until 1898 that a series of production delivery trucks was offered by the Winton Company, of which they sold 13 that year. Trucks were no overnight sensation though, and while growth was rapid in one sense, part of it was due to rapid increase in population and general business and industry in the United States rather than any big switch from those using animal power.

I can recall, as a boy in my Texas home town, that one of the local dairies still delivered their products to homes by means of a horse-drawn wagon as late as 1940. The City of New Orleans, Louisiana, still collected its garbage in the central City and downtown areas in mule-drawn wagons, until 1950; and also in New Orleans, hawker wagons, that is some selling produce door to door still used mule or horse-drawn wagons up to 1970. I took pictures of A. Willett's Sons' sugar factory in Plaquemine, Louisiana in 1949 where, at the time, sugar-cane was hauled from the fields to a small narrow-gauge railroad operated by coal-burning steam locomotives – hauled that is by ox-drawn carts. The local workers still rode around in wagons and carriages or on horseback at times.

Over in Mississippi, farmers who used mules to pull plows also carried their products to market or just went shopping in nearby towns in their mule-pulled wagon, and Mississippi highways were all built with wide grassy shoulders to accommodate such traffic, which was apparent well into the 1960s. Today a few religious sects such as the Amish still only get about by horse-drawn vehicles.

But in the '70s the motor truck did, indeed, dominate haulage, and it also marked new uses of the truck that increased sales beyond the wildest dreams of a few years before. Light trucks were bought because they were considered an 'in' or chic thing to do, that is drive to a country club or to an opening night in a fancy truck, rather than a passenger automobile. The young took to the small vans, which they made into travelling bedrooms complete with bars, television, stereo, as well as beds. But most common of all was the camper, or small motor home tacked on the back of a pickup truck, which gave many who never had the means for a country retreat, one of their own that they could take fishing, hunting, or just exploring. Such items caused truck production to burst over the four million mark in 1979 alone, in a ratio of one truck to every two and a half passenger automobiles produced.

But there were changes in the haulage trucks too. Early in the decade, an age of 'hot rods' showed up, equipped with huge diesel engines of 450 to 600hp that enabled a usual '18-wheeler' (tractor-truck and semi-trailer) to haul 30 tons down all types of highways at speeds of over 90mph when no one was looking, and legally in some places at speeds up to 80mph. But then somebody shut off the fuel supply and Americans found the pumps closed all of a sudden, with a new national maximum 55mph speed limit; prices for gasoline which in places had been 16 cents a gallon, were around the dollar mark at the end of the decade, so there was a big shift to longer-lasting, more fuel economical diesels, instead of gasoline (petrol) engines that had dominated before. Diesel engines, in spite of their huge first cost penalty, dominated in vehicles that also rose three times in price in the decade, causing down time and repairs to be examined more closely than in the past. Other fuel-saving items such as air foils and radial tires appeared in quantity too.

Creature comforts also became almost standard, such as air-conditioning most large trucks; things like automatic transmissions appeared in even large trucks in numbers unknown before 1975.

These revolutions, though, will not be apparent in the following pages to any great extent, as to see the vehicles does not tell all the story. You will, however, note that trucks of the seventies often were less streamlined than they had been years earlier, and had more boxy, less complicated designs, all of which is aimed at more reliability, and easier and cheaper maintenance. The selection that follows is only a scratch, though, on the over 200 brands of vehicles made in the period, and over 1,000 models available, but hopefully it will give a pleasant taste of what there was.

ELLIOTT KAHN
Clearwater, Florida
October 15, 1980

Airstream

Airstream is an old name in aluminum-bodied caravans, dating back over forty years, but a relatively new one in the truck field. In 1975 this company began to make delivery vans on forward-control chassis from Ford, Chevrolet or GMC, and also similar-looking motor homes and small buses under the name Argosy. A special line is in delivery vans for UPS (United Parcel Service), the largest independent delivery service in the United States.

1978 Airstream UPS van. This has much the same lines as the earlier UPS vans which were built for them by Flxible-Southern (Rohr), and described on page 46. From 1976 onwards, UPS vans were made mainly by Airstream, though UPS made a few themselves. All UPS vans are numbered, and the UPS-built units were the upper 59,000 series. This van, numbered 65,766, is therefore definitely Airstream-built.

American-General

American-General is the truck division of American Motors, but makes primarily military and postal vehicles, and also buses. Its trucks come in two distinct lines: small postal vans based on Jeep designs, but with two-wheel drive, and large 6 x 6 military trucks. The division did not really come into existence until 1971 when it took over a former Studebaker truck plant which had been purchased from Kaiser Jeep, and in 1970 it was decided that all government fleet sales would be handled by the new firm. In 1973 a new larger postal van with 6-cylinder AMC engine was introduced, being modified in 1974 and again in 1979, each time the sheet metal being simplified. The military trucks were continued, many being supplied to foreign countries, and in 1978 the firm obtained a US military contract to build over 5,500 trucks of much larger size than before, some up to 75,000lb GVW. In building these, AM-General co-operated with Crane Carrier Corp, using the conventional shape of the CCC Centaur model; CCC supplied many parts, but assembly was by AM-General. In 1977 a new line of electric Jeep Dispatchers was introduced, of which more than 500 were sold in the next three years. Production of these is planned to continue into the 1980s.

1979 American-General Jeep Dispatcher differs from the Jeep Corporation product in that it does not have four-wheel drive, but it is built in Jeep's Toledo, Ohio plant. All such units are sold directly to the US Postal Service only, but the Post Office resells them to the public after eight to twelve years use. These use the same engine as the $\frac{3}{4}$-ton shown on page 8.

1975 American-General postal van, built in the South Bend, Indiana plant that used to turn out Studebaker trucks. It has an AMC 231 CID ohv 6-cylinder gasoline engine developing 80hp.

1974 American-General M-35 2½-ton 6 x 6 military truck, the most common unit of this size in the United States military services. The design dates back to the 1950s, when it was built by several firms including Reo, Studebaker, GMC and International. When the Studebaker plant was sold to Kaiser Jeep in 1965 the name was changed to that of the new owners, this name continuing until 1970 when they started to be sold under the name American-General. Originally a 90hp Continental gasoline engine was used by Kaiser Jeep, this being replaced by a slightly more powerful AMC design. This type of truck was last built in 1978, the basic design having a production run of 27 years longer than that of the Model T Ford (19 years) or the Mack Model AC (22 years).

American Hoist

1971 American Hoist 8 x 4 crane carrier, made by a St Paul, Minnesota company dating back to 1928. Four- and five-axle chassis are marketed, some being made for American Hoist by CCC, Hendrickson, or Dart. This one has a Detroit diesel V-8 engine, but Waukesha gasoline or Cummins or IHC diesels are also available. Lift capacity of this Series 500 unit is 75 tons.

American La France

American La France as a unified name dates back to 1901, when several firms were merged, some of them dating from 1830s. From its horse-drawn days onwards the firm has been the largest supplier of fire apparatus in the United States. Motor trucks were first made in 1909, and their quality is such that some units built in the 1920s are still in use as back-up vehicles in 1979. American La France offer fire service bodies on chassis of other makes, but they are mainly known for their custom-built units, which are made in three series, Pioneer, Pacemaker and Century, in ascending order of price. Four-and six-wheeled chassis are offered, all with diesel engines by Detroit or Cummins.

1972 American La France 1000 Series pumper owned by the City of Sarasota, Florida. This was the top of the range ALF at the time, being supplemented by the Century line in 1973, which had a squarer cab shape. It has a Detroit diesel engine, though a few of this series were equipped with Allison gas turbines. They were by no means troublefree, and were phased out in 1974 after two years.

1977 American La France Pioneer II pumper putting out 1,250 gallons of water per minute, in service with the South Trail Fire Dept of Sarasota, Florida.

1972 American La France 900 Series pumper, painted red, white and blue to celebrate the Bicentenary of the United States in 1976. Open cockpit fire trucks, once common, are now pretty rare, but the mild climate of Vero Beach, Florida, where this one is in service, make such a design more practical than in most places. The 900 series was made from 1958 to 1974.

Autocar

1978 Autocar Series DC chassis will end up under a concrete mixer or perhaps a concrete block carrying body. With six-wheel drive this is a typical Autocar chassis, although they are largely custom-built trucks with special features to customers' order, so few are identical. This one has a steel hood and cab which dates back to the 1950s and was also used by Diamond-Reo on their C-116 construction trucks. Engines in 1970s Autocars were all diesels, either Caterpillar, Cummins or Detroit.

Autocar is one of the oldest names in the American industry, having built their first car in 1897 and their first commercial vehicle in 1899. Since 1953 Autocar has been owned by White Motors, though still having its own plant at Exton, Pennsylvania, until 1980 and its own designs. These are now mainly specialized vehicles for the construction industry, and all are conventionals since the dropping of the CK series in 1975. Autocars main plaint is at Exton, but some vehicles are built in other White plants such as New River Valley (Virginia), Ogden (Utah) and Quebec (Canada); in the latter plant a twin front axle dual steer chassis is built.

1972 Autocar Series CK half cab concrete mixer. This unusual design was based on a type developed by Cook Brothers of Los Angeles and made by them from 1956 to 1961. Autocar and Freightliner both took over the design and made it with their own front end patterns, but Autocar phased it out in 1975 because the relatively small radiator was unable to cool the increasingly large engines that customers demanded. Engines were mainly Cummins six or Detroit diesel V-8s from 230 to 375hp. The vehicle was available in two, three or four axle form, the one shown here having a non-driven tag axle.

1978 Autocar Construcktor trucks replaced the White Construcktor of similar construction in 1976, after White had built the model for ten years. The reason for the change was that Autocar were better geared than White to custom trucks of individual design to buyers' requests. Construcktors come in a variety of designs, with set-back or normal axles, two or three axles and wheelbases from 146 to 275 inches. The trucks illustrated have fiberglass hoods.

Bantam

Bantam is a make of mobile crane that originated in the 1940s when a small self-propelled railroad crane was built by the Ernest Scheid Corp of Waverly, Iowa. From the early 1950s truck versions were offered as well, with ratings of up to 30 tons lifting capacity. Bantam today is owned by the Koehring Company who also market larger brands of crane carrier under the names Bucyrus-Erie, Lorain and Koehring, as well as huge Dumptor trucks used in mining. Engines in Bantam cranes have been Ford, Chrysler or International gasoline units, or Cummins or Detroit diesels.

1975 Bantam Telekruiser 4 × 4 rough terrain mobile crane, powered by a 136hp Detroit diesel 4–53 4-cylinder engine, with Allison automatic power shift and top speed of 27mph.

1975 Bantam Model 426 6 × 6 mobile crane, available with either Chrysler or International gasoline, or Detroit diesel engines. The truck is rated at 49,000lb GVW. Five-speed Fuller transmission is standard, but a ten-speed one can be ordered with the diesel option.

Big Wheels

1976 Big Wheels Located in Pana, Illinois, this company is one of three in the United States who specialize in trucks with high flotation tyres for work in swampy ground. Engine and cab are GMC, but Big Wheels fit their own chassis and suspension as well as the oversize tyres. This particular truck is used in Florida for spraying oil on pools where mosquitoes breed.

Blue Bird

Blue Bird of Fort Valley, Georgia, are best known for their buses, which they have built since 1927 on other chassis, and from 1952 added their own chassis to their range. In 1962 they introduced a luxury motor home on the 35-foot bus shell, which they named the Wanderlodge. A number of these were fitted up as travelling sales rooms, libraries and classrooms or, as in the example shown, as a mobile blood bank. Engines were Ford, GMC or International gasoline or Caterpillar diesels. Blue Bird also made some six-wheeled bottle delivery vans for the Coca-Cola company in 1971, but these proved to be underpowered.

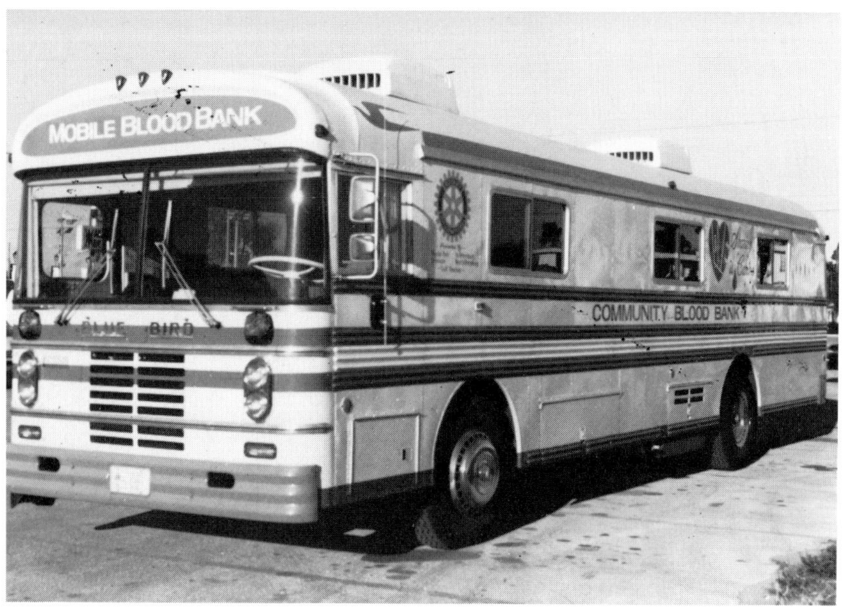

1976 Blue Bird Wanderlodge fitted out as a community blood bank. This example has a Caterpillar diesel engine and is fully air-conditioned. Prices of Wanderlodges run from $65,000 basic up to as much as $365,000 for a very specially equipped example.

c. **1970 Brockway Model 361 6 × 4 truck.** This was as large a rigid truck as Brockway built, with GVW up to 60,000lb. Rockwell or Eaton rear axles were used, with Hendrickson rear tandem suspension.

Brockway

Brockway was one of America's finest trucks when the firm went out of business in early 1977, victim of labour troubles and declining sales. While Brockway was called an assembled truck, they in fact rebuilt many of their components such as electrics to their own standard, modified engines and created their own conventional cabs, though they used Mack sheet metal on their cab-overs. The care they used to build a truck made it more costly, the major reason why the Cortland, NY firm was closed down by Mack who had owned Brockway since 1957. In the 1970s they made two series of cab-overs, the regular 400 and the low-profile 550, and also short and long-hood conventional models. Engines were mainly Detroit, Cummins and Caterpillar diesels, though a Continental gasoline unit was still available into the early 1970s if anyone wanted to order one.

1974 Brockway Model 360 tractor trailer unit, with auxiliary rear axle on the tractor, lifted from the ground in the photo. This axle was only used when the trailer was fully loaded, the tractor driving on the leading rear axle only. A semi-automatic Huskiematic transmission was an option on this model.

1974 Brockway Model 400 tractor trailer unit using Mack-designed cab supplied by Bethlehem Steel. With a choice of Caterpillar, Cummins or Detroit diesels, horsepower ranged from 230 to 434. This unit is rated at 35,000lb GVW, but tandem axle tractors could pull loads up to 73,000lb.

Caterpillar

Caterpillar of East Peoria, Illinois, is a well-known name in the field of tracked vehicles, and also as a maker of diesel engines and transmissions. They also make many other products including Towmotor industrial trucks and a range of heavy dump trucks, ranging from 60 tons to over 100 tons capacity. Engines vary from a 450hp V-8 to a 870hp V-12 in the 135 ton 777 model.

CCC

CCC The Crane Carrier Corporation began business in Tulsa, Oklahoma in 1946 by reinforcing surplus military chassis and selling them to builders of cranes. After a few years they found that the available chassis could not cope with loads greater than 15 tons, so in 1952 they built their first complete truck chassis. A year later they bought up the Available Truck Company of Chicago, an old-established maker of specialized chassis, and moved this operation to their Tulsa plant. They soon evolved a line of crane carriers with from two to six axles and capacities of up to 100 tons. Most of these had one-man cabs. More recently CCC have branched out into other truck fields such as chassis for concrete mixers, refuse trucks and conventional trucks for military or export sales made in conjunction with AM-General. A Canadian plant was operated from 1958 to 1979 which made some models specific to the Canadian market including logging trucks.

c. **1975 CCC Century M Series 6 × 6 concrete mixer,** typical of many used in construction industry.

1978 Caterpillar Model 773 dump truck with 600hp Caterpillar V-8 engine and a 50·6 cubic-yard capacity body. It has nine forward and three reverse gears.

1979 CCC low-profile crane carrier with Bucyrus-Erie crane. A choice of Cummins or Detroit diesel engines is mounted to right and rear of driver, while the cab is purchased from other suppliers, so looks almost identical to those of Hendrickson or Lorain crane carriers. Similar units are used on some fire service and airport crash trucks supplied by firms such as FWD and Oshkosh.

1975 CCC 8 × 4 crane carrier chassis, awaiting a crane by American Hoist which will be fitted by the dealer. This size chassis can carry a crane of up to 75 tons lifting capacity.

1975 CCC Centurion 6 × 4 refuse truck with front loading. Allison automatic transmission is standard on this model, and there is a choice of Caterpillar, Cummins or Detroit diesel engines from 237 to 310hp. Latest Centurions have a much lower profile cab with easier entry.

Chevrolet

Chevrolet is the largest producer of passenger cars in the world, and has held first or second position in the truck field too for many years. During the 1970s Chevrolet offered a wide variety of truck models, but was mainly prominent in the lighter end of the market. It was announced that the heaviest Chevrolet trucks would no longer be made after 1st October 1980, and indeed the heavy Chevrolets are in fact GMCs with Chevrolet badges anyway. In return the lighter models of GMC are made by Chevrolet. With the possible exception of the school bus chassis and the LUV pickup, Chevy offers nothing that is not available with different trim under the GMC label. The late 1970s range ran from the LUV (Light Utility Vehicle), which is actually a Japanese Isuzu with US-built pick-up body, to the huge Bruin and Bison conventionals and Titan cab-over trucks. Engines are mostly gasoline in the smaller models while the larger use diesels by Caterpillar, Cummins or Detroit, with perhaps more of the latter, which is understandable as it is a GM product.

1971 Chevrolet T-60 26,000lb GVW chassis fitted with fire pumper body. Built by GMC, these used Chevrolet-built engines, two sizes of six and three of V-8, up to 1977 when the sixes were dropped.

1974 Chevrolet Series G van, available in panel-sided van form as here, or as a minibus with seating capacity from six to fifteen passengers. This was one of the best-selling vehicles in the United States throughout the 1970s, with sales running from 200,000 to 300,000 per year. It was, and is available with either 4-speed syncromesh manual or automatic transmission.

1977 Chevrolet LUV pick-up has been marketed as a Chevrolet product since 1971, but is in fact made by Isuzu, of which firm General Motors owns some 34% of the common stock. In the latter 1970s a 4-wheel-drive version has been made. About 300,000 LUVs have been sold since 1971.

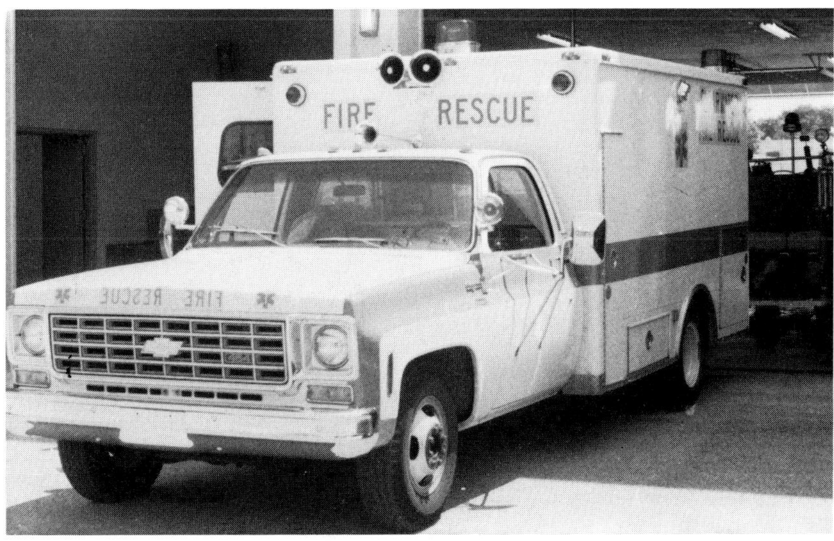

1976 Chevrolet C-30 with ambulance body by Southern Ambulance Body of La Grange, Georgia. This is a typical ambulance which has almost replaced the older sedan-based vehicle. The engine is the largest 454 CID V-8. (Gasoline)

1979 Chevrolet C–65, a typical conventional truck in Chevrolet's medium range, but with an unusually low trailer used for carrying beer. The usual engines in these trucks are either a 292 CID or a 454 CID V-8. (Gasoline)

1976 Chevrolet C–70 tandem axle 6 x 4, built by GMC at its Pontiac, Michigan plant. This particular truck is rated to haul a load of 60,000lb, which is 10,000lb over top factory specs, indicating that the owner either overloads his truck or uses larger than standard tyres or helper springs at the rear. This model could be had with either a Chevrolet V-8 427 CID gasoline engine, a Detroit diesel six 6—71, or a Caterpillar V-8 3208.

1979 Chevrolet Bison, a new heavy-duty modern design which is also marketed by GMC as the General. Some sixteen different diesel engines are offered, from Caterpillar, Cummins or Detroit, either in-line sixes, V-6s or V-8s. Two-and three-axle versions are made, with GVWs from 25,900 to 66,000lb. There was also a shorter variant called the Bruin.

1975 Chevrolet Titan tractor trailer unit which is identical to the GMC Astro 95 except for trim differences. The Titan changed very little in appearance from the first 1971 models to the final ones of 1980, which were announced in late 1979. They were available with the three leading makes of diesel engines in sizes from 200 to 390hp.

1980 Chevrolet Titan SS Series 90 which has a larger grille than the standard Titan, more de luxe features inside, fancier paint and, of course, higher price.

Coleman

Coleman dates back to 1922 when they followed Holmes, makers of all-wheel-drive trucks. Production was never large, but they were pioneers of all-wheel-drive oilfield trucks, and were also early users of oil engines – Hesselmans which used a spark plug, not diesels. In 1948 the company name was changed to American-Coleman. The major product of the 1970s has been the G–40 tractor, mostly used as an aircraft tug although it also finds applications in freight terminals. In 1977 Coleman obtained the marketing rights to the Champion yard spotter formerly made in Oklahoma and Texas.

Cushman

Cushman is often overlooked among commercial vehicle makers because of the small size of their products, but they have been in business since 1936, making a wide variety of three- and four-wheeled light trucks. All use air-cooled gasoline engines with one or two cylinders, from 18 to 22 hp. Load capacity of the largest is one ton.

*c.*1975 **Cushman 3-wheeled police truck** in service in Clearwater, Florida. Similar units with different bodies are used for sanitation service, for small fire pumpers in shopping malls, warehouses and factories, and for lawn maintenance.

1978 Coleman G–40 tractor with four-wheel-drive and steering, powered by a choice of Detroit or Cummins diesel engines, with power steering and Allison automatic transmission.

Custombilt

1979 Custombilt mobile truck crane, product of a relatively new small company from Kansas City, Missouri. Dating from 1976, Custombilt make three models of truck crane, or elevated platform unit. Standard engine is the German Deutz air-cooled diesel, but Caterpillar, Cummins or Detroit can be had as well. Sizes range from 13,800 to 36,820lb GVW, and booms vary from 25 to 85 feet high.

Dart

Dart is the oldest company in what today is known as the Paccar group, whose better-known truck products are Kenworth and Peterbilt. Today Dart makes mainly heavy off-road dump trucks, but they made a wide variety of road-going trucks in earlier years, dating back to 1903. At present located in Kansas City, Missouri, Dart also make crane-carrying chassis for American Hoist and Derrick, and also large aircraft refuelling trucks.

1979 Dart Model 3100 100-ton dump truck. Unlike rivals such as Lectra-Haul and the larger Terex and Wabco models, Dart use mechanical final drive in preference to electric drive, as maintenance is easier and weight lower. This model has a choice of Detroit diesel or Cummins power units, both V-12s developing 1000 and 1050hp respectively.

Diamond-Reo

Diamond-Reo resulted from the 1967 merger of two fine old names in American trucking, Diamond T and Reo. Conventionals and cab-overs were made, many tractor trailer units although there was also a military 6 x 6 similar to the former Reo design, and an unusual refuse truck with twin cabs resembling telephone booths. Unlike most big truck makers, Diamond-Reo built their own gasoline and LPG engines into 1974, while also offering diesels from the leading makers as well. Even Chevrolet gasoline engines were offered in the small Trend series of cab-overs. In 1974 the firm was declared bankrupt and after prolonged negotiations the name and designs were acquired by a dealer, Osterlund Inc of Harrisburg, Pennsylvania. They put one model into limited production in 1978, and this continues today.

c. 1972 **Diamond-Reo C–101 6 × 6 concrete mixer truck.** Powered by a Detroit diesel engine, this is a successor to the 6 × 6 models made by the former Reo company.

1971 Diamond-Reo Royale cab-over tractor trailer unit with sleeper cab. This model dated from 1967 and was a continuation of the earlier Diamond T design with changed grill and trim. When the Royale II was introduced, this model was retrospectively named the Royale I.

1973 Diamond-Reo C–116 6 × 4 tractor with flat bed trailer.

1974 Diamond-Reo C–119 Raider tractor, introduced in mid-1974 and only made for eight months, so a rare sight on the roads. The bonnet was fiberglass and lifted forward in one piece, over the grille section for servicing.

1979 Diamond-Reo C–116 6×4 chassis, one of the Harrisburg – built DRs which are still in production. All models so far are 6-wheelers, with either 6×4 or 6×6 drive and Detroit or Cummins power. This photo was taken at the Harrisburg plant of Osterlund Inc; the trucks are brand new and awaiting registration.

Divco

1970-type Divco 300 series delivery van powered by a 170hp Ford six engine. Though this example was made in the 1950s it is identical in appearance to early 1970s Divcos. It can carry a load of three tons, and the twin rear wheel models have a load capacity of 4½ tons.

Divco was named from the original builders, Detroit Industrial Vehicle Co who introduced a door-to-door delivery truck in 1925. This has been the main product of the company ever since, and the round-nosed design introduced on 1938 models is still recognizable in today's Divcos. The company has had many different owners in its lifetime (two in the 1970s) and production fell at one time to as low as seven units in one year. Most Divcos have gasoline engines supplied by Ford, and earlier Continental and Hercules, though a few diesels have been used as well.

Dodge

Dodge has been a division of Chrysler Corporation since 1928, and at the beginning of the 1970s offered a complete range of trucks from a half tonner to units rated at 76,000lb GVW, including conventional and cab-over models. However at the end of 1975 all trucks larger than 30,000lb and all cab-overs except the small vans were dropped, and two years later even the medium-sized trucks were dropped, with the exception of a 16,000lb chassis sold almost entirely to builders of motor homes and buses. Current production, therefore, is centred on vans and pickups only.

1975 Dodge D-300, usually referred to as a one-ton model, though with heaviest options it can haul a two-ton load. This one has an ambulance body for the Largo, Florida, Fire Department, and is powered by a 440 CID V-8 giving 235hp.

1974 Dodge/W-300, similar to the D-300 but with four-wheel-drive. Although Dodge-built, this fire truck carries the name plate of Pierce, after the Appleton, Wisconsin, firm which makes the fire apparatus. This is what is known as an attack truck, having a 300-gallon per minute pump and also carrying about 200 gallons of water in its own auxiliary tank.

1972 Dodge CN-900, typical of the heavier Dodge series from the early 1960s until 1975 when they were last offered. Of similar appearance was the CT-800 designed for semi-trailer work. Both had either Cummins or Detroit diesel engines under the bonnet, but some of the 800s were also offered with Dodge six or V-8 gasoline engines too.

1971 Dodge D-500 tipper truck, typical of medium-sized trucks owned by municipalities in the 1970s. It was available with an in-line six or V-8 gasoline engine.

1973 Dodge Big Horn tractor trailer unit. This was the largest Dodge truck made, and was listed only from 1973 to 1975. Possibly fewer than 500 were ever made, and they are a rare sight. GVW was in the range 59,000 to 75,000lb, and engines from all the three well-known diesel makers were offered, plus the lesser-known Allis-Chalmers Big AL, only one of which was actually made.

Duplex

Duplex is a relatively unknown brand of truck even to Americans, yet its history dates back to the early years of the century, and a four-wheel-drive truck was built in the 1907–09 period. In the early 1950s production dropped to five units per year but the company was rescued by Warner & Swasey, a Cleveland-based manufacturer of tools and light cranes and the Gradall, which was a truck-mounted digging machine for clearing land and laying cables. W & S built new Duplex chassis to carry the Gradall, a line of crane carriers and fire apparatus, and within a few years Duplex production was up to 900 units per year. In the 1970s Duplex offered three types, a 4 × 4 dump truck (discontinued in 1974), fire apparatus and crane carriers. The latter were renamed Badger in 1976, and late in 1977 the fire engine manufacture was taken over by the Nolan Company of Columbus, Ohio, and production moved to a new plant at Midvale, Ohio, where about fifty chassis are made per year. At about the same time Warner & Swasey sold the Badger division to new interests in Chicago, though the plant remains at Winona, Minnesota.

1977 Duplex fire engine chassis. This looks like many other brands of American fire engine as it uses the Cincinnati cab made by the Truck Cab Co of Cincinnati, Ohio, and supplied to at least twenty makers of fire apparatus. Usual power plant is a Cummins or Detroit diesel.

1973 Duplex-based Warner & Swasey Gradall unit, a model which could also have been supplied to run on rails or with a crawler tractor base. The truck unit normally has a Detroit diesel engine, while the hydraulic crane has its own power unit, usually a 120hp IHC six, though a 3-cylinder Detroit diesel was an option. The example shown was photographed in Switzerland.

EVO

***c*.1972 EVO Model 1650 front drive refuse truck.** This is a product of Lodal Inc of Kingsford, Michigan, and represents a novel approach to the problem of refuse collection and disposal. The EVO can pull or self-load several small containers onto its back in house to house collection work, and when it is full it goes to a transfer yard where its load is lifted onto a much larger truck for taking to the dump site. It has left- and right-hand steering. This particular truck has Chevrolet chassis and engine, but other EVOs made in the 1970s have used Ford, Dodge and IHC components.

Fiat-Allis

1979 Fiat-Allis Model 161 tractor scraper. Fiat-Allis was formed in 1974 as a combination of Allis-Chalmers, makers of scrapers and many kinds of construction equipment, and Fiat of Italy who owned two-thirds of the new corporation, with Allis-Chalmers having the other third. Many of the designs came from Fiat, though the engines were a continuation of Allis models which themselves derived from the old Buda designs, as Buda had been acquired by Allis in the 1950s. This particular model uses a Cummins diesel engine.

FMC

Flxible-Southern, *see* Rohr (page 46)

Ford

Ford, like Chevrolet, is a universal provider to American truck users, making an enormous variety of vehicles from light pick-ups to the CL-9000 tractor trailer unit for a maximum of 80,000lb GVW. Although the 1970s have seen several important new Fords at the top of the range, their largest selling models are the lighter ones, especially the pick-up which is the largest selling vehicle on the US market, outselling the nearest passenger car model by two to one. (Chevrolet pick-up is the second best seller.)

While many consider Dearborn, Michigan, to be the home of Ford, it is many years since any Ford trucks were built there. All the heavy models come from the Louisville (Kentucky) plant, while other Ford truck plants are located in Kansas City (Missouri), Norfolk (Virginia), St Paul (Minnesota) and San Jose (California).

1972 Ford Econoline ambulance. This is an ambulance and rescue version of the famous Ford van which has been made in a variety of forms since 1968. It has a Ford V-8 gasoline engine; GVW is 10,000lb.

1978 FMC fire engine. These initials stand, perhaps rather surprisingly, for Food Machinery Co, a conglomerate owning a wide variety of firms including Wayne (street sweepers), Link Belt (conveyor systems and truck-mounted cranes) and John Bean (fire apparatus). They also make their own rear-engined chassis for motor homes and buses. This FMC fire engine has a chassis by Spartan Motors of Charlotte, Michigan with FMC pumps and other equipment.

1980 Ford Econoline van, of similar appearance to 1978/79 models.

This one is in the brown livery of United Parcel Service, the largest truck shipping line in the US, and is unusual because of its raised roof section normally found on camper units or minibuses. This is to provide standing room inside the van. Note the oversize tyres and wheels allowing a rating of 11,000lb GVW.

1975 Ford C-800 fire engine. Another C-line fire engine, on a much longer wheelbase, with snorkel unit by Pierce of Appleton, Wisconsin.

1978 Ford C-line fire engine. Ford's C-line is one of the all-time greats in truck design, as well as being the top selling cab-over in the world.

Introduced in 1956 it was the first tilt-cab Ford, and the design has been used by other manufacturers, notably Mack and FWD. This small fire engine bears the badge of Emergency One who make the fire equipment, as well as that of Ford.

1975 Ford L-9000 tipper truck. Made in a new plant at Louisville, Kentucky, the L series was introduced in 1970 and was the largest conventional line of Ford trucks. Diesel engines were usual, though gasoline options were offered in all models. This Caterpillar-powered tipper has a tandem powered axle with a third idler axle used only to carry heavy loads over lightly rated bridges.

1976 Ford LTS-9000 refuse truck, an example of the L series with set-back front axle designed to give a shorter wheelbase. Appearance is considerably altered, though in other ways they are similar to the regular L series, with the same engine options.

1979 Ford CLT-9000 tractor trailer unit. Introduced in 1978, this is the largest Ford truck model, and enables the company to compete with specialist heavy truck makers such as Kenworth and Peterbilt. Among its features are air suspension, even air ride for the cab alone that divorces it from the action of the frame. It is available only with diesel power, by Caterpillar, Cummins or Detroit, with horsepower ratings as high as 600. GVW is 80,000lb, the highest permitted in normal operation on US highways, though the truck could handle more if permitted.

1975 Ford LTS-9000 eleven axle unit. This centipede-like truck and trailer is typical of the multi-axle trucks used in Michigan where there are restrictions on the weight loading on any one axle. Thus a 12-axle combination may be needed for a load which could run on six axles in California. Though chiefly associated with Michigan, these centipedes are found in parts of Wisconsin and in Canada.

Freightliner

1975 White Freightliner WFLT tractor trailer unit. This truck is typical in having a tandem rear axle, though two-axle models were also made.

Freightliner began as the result of an operator's dissatisfaction with existing trucks. In 1936 Consolidated Freightways Inc modified some GMC tractors they were operating, and two years later brought out their own design of cab-over tractor which they named the Freightliner. It used more aluminum in cab and frame than was common at that time. After World War II they began to sell to other operators, and in 1951 they entered into a sales agreement with White who were to sell Freightliners alongside their own trucks. This led to the name White Freightliner appearing on the trucks, the name lasting until 1976 when the trucks became simply Freightliners once more. The original plant was in Salt Lake City, but post-war Freightliners were made in Portland (Oregon) and more recently additional plants have opened at Chino (California), Mount Holly (North Carolina) and Indianapolis. There is also a Canadian plant at Vancouver.

1979 Freightliner WFT-8164 tractor trailer unit. This is the less common conventional model of Freightliner introduced in 1974; all previous models had been cab-overs.

FWD

FWD began life as the Four Wheel Drive Auto Co of Clintonville, Wisconsin, making their first 4 x 4 truck in 1912. These were widely used by the US Army in World War I, and their successors played a big part in the second conflict, from 1939 to 1945. By 1960 the name was not entirely appropriate as the company was making some trucks with six-wheel drive, so the name was changed to the FWD Corporation. Later a highway tractor with 6 x 4 drive was offered, though the main product of the company has continued to be 4 x 4 and 6 x 6 chassis for concrete mixers, also oilfield trucks and crane carriers. The regular highway tractors were discontinued in 1974. FWD also makes Seagrave fire engines.

1970 FWD C-type 6 x 6 concrete mixer with Detroit diesel engine and 10 cubic-yard mixer. There were also chassis with four or even five axles.

*c.*1972 **FWD CB-66 Tractionmaster.** This 6 x 6 truck has a snorkel like a fire engine, but it is in fact operated by a power company to install telephone poles or to work on the lines.

*c.*1972 **FWD ForWarD Mover 6×4 tractor,** one of the last regular commercial trucks offered by FWD. Despite being a high-quality product, the makers could not compete with rivals like Kenworth, Peterbilt and Mack, and wisely decided to concentrate on their specialities such as all-wheel-drive construction industry trucks, crane carriers and fire engines. Both sleeper and non-sleeper cabs, as here, were offered.

GMC

GMC dates from 1911 when the General Motors Truck Co was formed from the merger of three firms, Reliance, Rapid and Randolph. Electric vehicles were also made up to 1916, and in the 1920s GMC grew to become one of the most important makers of commercial vehicles in the United States, adding the Yellow Coach line of buses and coaches in 1925. During the 1970s GMC has become increasingly a badge-engineered truck, with nearly all models being offered under the Chevrolet name as well. The smaller GMCs are made by Chevrolet, and the larger by GMC but are also sold under the Chevrolet name. The only exceptions are bus chassis and the front-wheel-drive motor home which was discontinued in 1978. GMC-built diesel engines were dropped in 1972, although the firm used Detroit diesels which are made by another division of General Motors. Gasoline engines used in smaller GMC trucks come from Buick, Chevrolet, Oldsmobile and Pontiac.

Gerstenlager

1974 Gerstenlager travelling library. Located in Wooster, Ohio, Gerstenlager is an old-established company dating back to the 1840s that today specializes in travelling libraries, though it also makes mobile post offices, sales rooms, classrooms etc. In addition, it makes large rescue vehicles for fire departments. Most Gerstenlagers are based on International running units.

1975 GMC P-3500 step van, with a Union City body. Similar-looking models with single rear wheels known as the P-1,500 and P-2500 are also made, while a P-4500 with 17,000lb GVW was added in 1977. There are parallel Chevrolet models and all are made in the Chevrolet plant at Detroit, Michigan.

1977 GMC 6000 Series with beverage body. With a rating of 24,000lb GVW, this unit is the largest of the 6000 series. Standard power was a Chevrolet or Buick-built gasoline engine, though a Detroit diesel V-8 was an option. Trucks of this type were the main reason why GMC ranked fourth in sales of medium and large trucks in the USA.

1975 GMC Astro 95 tractor trailer unit. This was the top offering in the GMC range and has been made since 1969. Always diesel powered, it comes in two-axle and tandem drive three-axle form, with a top rating of 80,000lb GVW and power outputs of up to 450hp. A. O. Smith makes the frames and Sheller-Globe the cab. Suspension on tandem models is usually Hendrickson, through Reyco has been used as well. About 100,000 of this model has been made, this figure including the almost identical Chevrolet Titan.

1978 GMC Brigadier chassis. Largest of the conventional GMCs are the General and Brigadier series introduced in 1977, the former with 108in BBC measurement, and the latter a shorter 92$\frac{3}{4}$in. The Brigadier came in two ranges, the 8000 and the 9500 (illustrated), and had GVW ratings from 23,000lb for the two-axle model up to 56,000lb for the top three-axle units. Gasoline engines are offered in the 8000, as were diesels from all three major makers, but the 9500 series came only with diesels.

1979 GMC Astro 95 tractor. This is one of the tandem axle Astros, and is fitted with a sleeper cab and Uniroyal Air Deflector above the cab. Increasingly popular all over the world in the fuel-conscious 1970s, air deflectors cut fuel consumption by as much as 10 to 15%.

Grove

1975 Grove four-axle mobile crane. Made in Shady Grove, Pennsylvania, Grove is one of the largest builders of mobile cranes and fire towers in the United States. Some of their chassis are made for them by other firms such as FWD, but it is thought that the larger chassis are Grove-built. These run up to 200-ton models on six axles. Power units are mostly Detroit, with Cummins as an option.

Hagie

*c.*1970 **Hagie light refuse truck.** These curious little vehicles are designed to operate in conjunction with larger refuse trucks in the same manner as the EVO (page 22), and are particularly useful in rural areas with poor roads, or in narrow alleys. They are powered by a rear-mounted 4-cylinder Ford Industrial engine and use hydrostatic drive. Made in Clarion, Iowa, the Hagie has, since 1977, been marketed by the Heil company and now carries the Heil nameplate.

Hahn

*c.*1970 **Hahn fire engine.** Like many fire engine makers in the US, Hahn use cabs by the Truck Cab Equipment Co of Cincinnati, Ohio, popularly known as the 'Cincinnati cab', which gives all such fire engines a very similar appearance. Hahn is an old-established company dating back to 1908 who made regular trucks up to the early 1950s, afterwards concentrating on fire engines and bodies for mounting on other truck chassis. Production of fire engines does not exceed 10 to 25 units per year today. Detroit diesel engines are mostly used.

P and H (Harnischfeger)

1977 P and H T-300 30-ton mobile crane. Harnischfeger of Milwaukee, Wisconsin, is one of the largest makers of mobile cranes in the United States, varying from small two-axle machines for rough terrain of 15 tons lifting capacity up to the huge six-axle T-1300 with 300 tons capacity. This T-300 is powered by a Cummins V-8 which drives the truck and powers the crane. Top speed on the road is 50mph.

Hendrickson

The name of Hendrickson is the best known for the tandem-drive axles which they pioneered in 1926, and which are widely used by truck makers such as GMC, Freightliner and International. However they have made trucks of their own since 1913, and have also built trucks for other firms, especially International for whom they made the large six-wheelers of the 1930s and more recent fire engines. Though small in output, the 1970s range of Hendrickson vehicles was as wide as that of any US make, including over-the-road tractor trailer units, crane carriers, bus chassis, aircraft refuelers, off-highway mining trucks, oilfield trucks and even a locomotive which could run on rails or road.

*c.*1972 **Hendrickson H-3 armoured truck.** Brinks armoured trucks for carrying money or valuable objects are a familiar sight in the US, but Hendrickson is a relatively rare chassis for them to use. This particular one was used to carry funds between Federal Reserve banks in Jacksonville, Tampa and Miami (Florida). The current H-4 is still much the same design as this H-3, and the H series dates back to the mid-1950s.

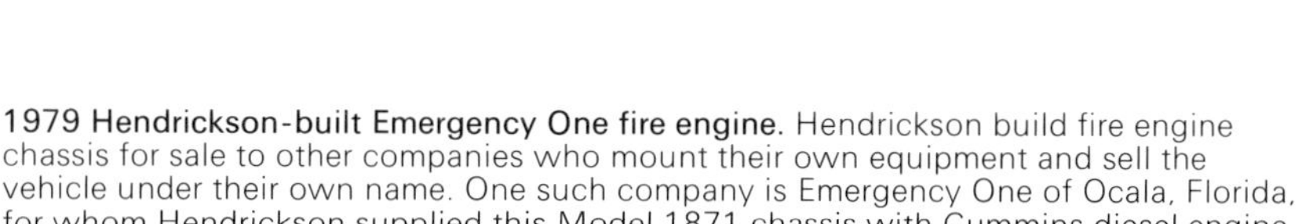

1979 Hendrickson-built Emergency One fire engine. Hendrickson build fire engine chassis for sale to other companies who mount their own equipment and sell the vehicle under their own name. One such company is Emergency One of Ocala, Florida, for whom Hendrickson supplied this Model 1871 chassis with Cummins diesel engine.

1978 Hendrickson H-4 dump truck. This is the most recent of the H series of heavy duty trucks for construction and commercial freight industries. Cab and bonnet are bought from International, and early H series had International engines as well, though these are now Detroit or Cummins diesels.

1979 Hendrickson Model 1000-D aircraft refueler. These twin-steer units are a familiar sight at many airports in the United States, some being made by Hendrickson and others by Dart. This one has a Detroit diesel 6V-53 engine developing 175hp and is suitable for fuel tanks of 8,000 to 10,000 gallons.

Ibex

Ibex is a little known builder of custom trucks located in Salt Lake City, Utah. The firm began in 1964 by making all-wheel-drive trucks for heavy duty work both on and off the highway, but the 1970s saw them enter the yard jockey market with the Flexi-truck which was also marketed for them by Flexi-Truc Inc of Secaucus, New Jersey. These are powered by a choice of Ford gasoline engine, or diesels by Detroit, Caterpillar or Cummins, and all have a hydraulically-operated rising fifth wheel. This is a Flexi-Truc F-2 of 1979.

International Harvester Company

The International Harvester Company was founded in 1902 from a merger of several prominent farm equipment makers. Their first truck was the Auto Wagon of 1907, a high wheeler type of vehicle popular with farmers, and larger trucks appeared from 1915. The first hundred Auto Wagons were built in Chicago where the company headquarters were, and are, but all subsequent production was at Akron (Ohio). More recently, other plants have been opened at Springfield (Ohio), Fort Wayne (Indiana) and, at the end of the 1970s, at Waggoner (Oklahoma) for extra large trucks. There was also a plant at Doncaster, England, which operated from 1965 to 1969. International make the most of the gasoline engines used in their light to medium trucks, and also diesels in the medium-sized field, buying the largest diesels from outside makers. Their range is very wide, from the Scout which is a Jeep-type 4 × 4 vehicle up to the Transtar tractor trailer unit for GVWs up to 80,000lb.

1974 International 1300 D 4 × 4 light fire engine, with Reading body. Apart from the Scout this was the smallest size of International, made also in two-wheel drive form as the 1000 and 1100 series. Standard factory bodies included station wagon and pick-ups with ordinary cab or crew cab for six passengers.

1977 International Binder with high capacity van body for carrying mattresses. This was a typical medium-sized International as made throughout the 1970s, rated at 17,000lb GVW and with a gasoline engine.

1977 International Binder 1600 tractor, built on an especially short wheelbase to meet Florida state regulations for pulling mobile homes on the highway. This load is 12 feet wide and requires a special permit. Disc wheels as here were not so common as the spoked type shown above.

***c.*1972 International CO-1600 Cargostar.** This cab-over design dates back to the mid-1960s and indeed is not all that different from the first International tilt cab made in 1954. Usually gasoline engines were fitted but diesels were an option.

1976 International CO-1700 Cargostar. The successor to older model which it replaced in 1975. It comes in various models from 1600 to 1950, and with six-wheel tandem rear it can be rated as high as 54,000lb GVW. A V-8 IHC gasoline engine is the most common, but you can get either an in-line six or a V-8 diesel as well, the 466 type which is also used in British Seddon-Atkinson trucks. Seddon-Atkinson has been owned by International since 1974. The Cargostar is still made in 1981.

1978 International S Series, one of the best-selling new designs of the late 1970s. The heavier models first appeared in 1977, and the later ones replaced the entire medium-sized line of 1600, 1700 and 1800 series in late 1978. The S Series, named after Springfield, Ohio, where most of the units are built, can have gasoline, diesel or LPG engines, all made by IHC, or diesels by Detroit, Cummins or Caterpillar.

1978 International Paystar 5000 Series contruction truck, with tandem axle drive and an unpowered tag axle which can be raised, as here, when the truck is lightly loaded. First introduced in 1972, the Paystar series can be had in 6 x 6 form or as a tractor trailer unit when a GVW of 180,000lb is possible, though only for off-highway work. Paystars are made in Fort Wayne, or in the new heavy vehicle plant at Waggoner, Oklahoma. Engines are a 228hp Detroit diesel six as standard, with the option of Cummins or Detroit V-8s up to 350hp.

1975 International CO-4070 Transtar II tractor trailer unit for nationwide furniture moving. The Transtar is International's offering in the highway tractor market since the mid-1960s, and comes with or without sleeper cab. The sleeper cab has a BBC measurement of 73 inches, which is the lowest in the industry, but it makes for rather cramped conditions. Unlike other Internationals, the Transtar comes only with diesel engines, by Detroit, Caterpillar or Cummins with a maximum of 450hp. The box on the top of the cab is the air-conditioner now mandatory on all long distance heavy trucks.

Jeep

Jeep is now the correct brand name for the light vehicles made for so many years, at first by Willys Overland, then by the Kaiser Jeep Co, and since early 1970, American Motors who renamed the firm Jeep Corporation. Still built in Toledo, Ohio, where the first Willys Jeeps came from in 1941, the Jeep is now made in a variety of forms including station wagon, pick-up or chassis and cab for mounting with a special body. All jeeps of the 1970s have had all-wheel-drive, and engines of four, six and V-8 layout are available, all gasoline fuelled.

1973 Jeep J-4000 pick-up, the first to be fitted with the new Quadra-Trac four-wheel-drive system.

Kaiser Jeep

Kaiser Jeep. This brand name was applied not only to the light 4 × 4 vehicles but also to heavy 6 × 6 military trucks built in the former Studebaker plant which was acquired by Kaiser in 1964. From then until 1970 the trucks were called Kaiser Jeeps, after which the acquisition of the plant by American Motors led to a change of name to American-General, by which their descendants are still known today. (See page 7)

*c.*1970 **Kaiser Jeep 6 × 6 ex-army truck.** These trucks were never available new to civilian buyers, but after a few years' military service many were sold off to various private customers. This one was acquired by Polk City, Florida, who have used it as a fire engine for combating brush fires.

Kenworth

Kenworth Truck Company was formed in 1923 by H. W. Kent and E. K. Worthington (hence the name) and became the most successful of any truck made in the Northwestern United States, their factory being in Seattle, Washington. After World War II Kenworth was acquired by the Pacific Car & Foundry Co, now known as Paccar Inc, and also owners of Peterbilt. Though Kenworth once built light door-to-door delivery vans and buses, they were known in the 1970s only for heavy trucks, some about as large as any produced in the world. The largest were oilfield and logging trucks not intended for highway use, with GVWs in excess of 220,000lb.

1975 Kenworth C-500 6 × 6 dump truck. This was a popular truck for the construction industry, being available in 6 × 4 or 6 × 6 form, with extra non-driven auxiliary axles if needed. Standard engine was a 290hp Cummins but diesels of all three major makers were available. Spicer or Fuller manual transmissions or Allison automatic were options, with axles from Eaton or Rockwell-Standard. An even larger version is known as the Brute; this has a steel and fiberglass cab in place of aluminum, and a heavier frame.

1978 Kenworth L-700 6 x 4. This model is made in Kenworth's Canadian plant at Quebec, alongside the Peterbilt 310 which is the same truck apart from trim differences. It is designed for sanitation work and local delivery of goods and beverages, having a small turning circle and a BBC measurement of only 58 inches. The only model is the three-axle tandem drive illustrated, but engines can be Cummins or Detroit diesels from 230 to 350hp.

1975 Kenworth K-100 cab-over tractor trailer unit. This is the cab-over version of the W-900 and is equivalent in every way, also dating from the late 1950s. Engine options are similar, and the tractor comes in two- or three-axle form. This large moving van is operated by Passport Transport who specialize in carrying antique and classic cars.

1976 Kenworth W-900 conventional tractor. This familiar shape of Kenworth has been made for over twenty years, and bears a distinct family resemblance to the radiator of the immediate pre-war Kenworths of 1940. Fitted with lightweight components such as cab and wheels of aluminum alloy, the W-900 has an excellent power-to-weight ratio with engines up to 450hp. It is popular with owner operators who often make use of the colour options available from the factory, or design their own colour schemes. While most units are six wheelers as here, it can be had with two axles as well. The six wheeler can be rated as high as 120,000lb GVW.

Lorain

1975 Lorain 17-ton mobile crane, one of the smallest in a wide range made by the Lorain Company which has plants in South Lorain (Ohio) and Chattanooga (Tennessee). Larger units run up to a lifting capacity of 250 tons and five axles. Lorain was probably the first company in the world to make a half-cab truck, which they did in 1939 with their first complete mobile crane. Earlier products had been mounted on other makers' chassis.

Mack

Mack is probably the best-known make of truck in America, possibly in the whole world. This firm entered the vehicle business with sightseeing buses in 1902, making their first trucks three years later. The most famous and long-lived model was the $7\frac{1}{2}$-ton chain drive Model AC, familiarly known as the Bulldog, and made from 1916 to 1938. Buses were made from the mid-1900s to 1960, and large off-highway dump trucks were added to the range in 1941 and were made up to 1979. Among other specialized vehicles are fire engines, of which Mack is the second best selling make, after American La France. They probably make a higher proportion of their own components than any other heavy truck builder, yet even so they offer engines by Scania, Caterpillar, Cummins and Detroit as well as their own diesels; they make their own transmissions but also offer Fuller or Spicer units.

The main Mack plant is at Allentown, Pennsylvania, but there is an important West Coast operation at Hayward, California which makes Mack Western models, and a Canadian assembly plant at Oakville, Ontario. Foreign assmbly plants are found in Venezuela, Pakistan and Australia. The Hagerstown, Maryland, factory which made off-road dump trucks now makes transmissions.

1974 Mack MB refuse truck, with a tilt cab design that dates back to 1956 and was not phased out until 1978; it was Mack's first tilt cab. Early units had gasoline engines, but during the 1970s most had Mack-built diesels though Scania and Detroit engines were fitted to a few as well.

1979 Mack Interstater, the top model of the F series first introduced in 1978. It has a tilt cab with a number of lightweight components of aluminum alloy and fiberglass. This is the 'dressed up' version designed to appeal to the owner driver, with more plush interior.

***c.*1970 Mack DM 6×6 concrete mixer.** 'DM' stands for Dump and Mixer, for which tasks this chassis was designed. It was made from the mid-1960s to mid-1970s, with little change. There were many variations of this model, which was largely made to customer's order. This one has all-wheel-drive so is a DMM model, though 6×4s were made as well.

*c.*1978 **Mack Western RS-700 6 × 4 chassis.** This is one of the Western range made at Hayward, California, and can be identified as such by the 'collar' around the grill, and by the mesh grille itself. Lightweight components are standard, so these trucks weigh as much as 500lb less than similar Eastern-built models.

1977 Mack fire engine with Mack's own fire tower unit, their answer to the more popular Snorkel brand offered by most other builders. This demonstrator carried a $140,000 price tag in 1977, but the price is now over $180,000 for basically the same vehicle. It has an hydraulically raised platform with water tower on it that can reach over 90 feet in the air.

1974 Mack Type R tractor trailer unit. Made since 1966, the Type R is one of the most widely sold of all Mack models. It comes in twin- and three-axle form, as a rigid truck and a tractive unit, and usually with Mack diesel engine, though a few gasoline units were made in the early 1970s. The fiberglass bonnet tilts forward for servicing. GVW for the R-401 starts at 27,000lb and goes up to 52,000lb on the tandem drive R-785, while tractor models are rated for up to 80,000lb.

*c.*1975 **Mack Type F-700 tractive unit.** This is the cab-over version of the Type R and like it comes in two- and three-axle forms. This sleeper cab model has an 86½-inch BBC measurement, against 56½ inches for the ordinary cab. A choice of engines from 237 to 430hp is available with Mack and outside brands available, though Caterpillar was not added to the list of options until 1977.

1977 Mack Western WL 700 LS tractor trailer unit, popularly known as the Cruiseliner. Wheelbases (all units are tandem drive) range from 146 to 254 inches. Mack, Cummins or Detroit diesel engines are available. The Cruiseliner has a larger radiator and cooling system than the older F Series.

Marmon

Marmon is a name mostly associated with a high-quality passenger car made in Indianapolis from 1902 to 1933, but it is also that of an equally high-quality truck that has been made in Garland, Texas, since 1964. Marmon trucks are large highway tractor trailer units powered by the usual choice of diesel engine. The firm made about a thousand trucks during the 1970s, but were badly hit by the mid-decade depression and sales dropped to about 50 per year, causing loss of most of their dealer organization.

1975 Marmon CHDT highway tractor, the conventional version of the HDT, which was added to the range in 1972, and offered the same engine options. Note the twin vertical exhaust stacks and air-conditioner unit at the side of the hood.

1973 Mack M-75 SX 75-ton end dumper. One of the range of outsize off-highway dump trucks that Mack made up to 1979, the M-75 SX had a tandem-drive-axle and was powered by a 700hp Cummins V-12 engine of 1710 CID. Alternative power units were a V-16 Detroit diesel or 800hp V-12 Cummins. The complete weighed 105,000lb, stood over 12 feet tall and was 13 feet wide.

*c.*1970 **Marmon HDT series tractor trailer unit.** This is the first type of Marmon truck, introduced in 1964 and made with little change up to the mid-1970s, when a larger radiator was adopted. While a two-axle model is listed, most are three axle as pictured. This well-worn example has lost the lower part of its radiator grille as well as the bumper.

1978 Marmon HDT series tractor trailor unit, similar to the above example, but with larger grille. Engine options in these Marmons were either Cummins or Detroit diesels, with Caterpillar being offered as well from 1977. Current ratings are from 250 to 525hp. Sleeper cabs of 86- to 110-inch BBC may be chosen.

1979 Marmon Transmotive road/rail truck. This unusual vehicle is made by a descendant of the old Marmon Herrington company, located in Knoxville, Tennessee, and has no connection with the Texas-built Marmon trucks. It is designed to take railroad maintenance crews to remote locations, hence its ability to run on roads as well as rails. Power is by an American Motors 360 CID V-8 and there is a Chrysler four-speed automatic transmission. GVW rating is 12,000 to 14,000lb.

Master Truck

1978 Master Truck 6×4 refuse truck, product of a California company which specializes in sanitation work. The Master Truck was introduced in 1972 and comes in two- or three-axle form, with left- or right-hand drive and a choice of all three major diesel engines, plus the British Perkins and German Deutz air-cooled diesel unit. About 600 Master Trucks have been made so far.

Maxim

Maxim is one of the oldest names in American fire apparatus, beginning by making ladders in 1888. Motor fire engines came in 1914, and their first complete chassis was built two years later. Located in Middleboro, Massachusetts, Maxim have gone quietly on making fire engines down the years, tractor units as well as two- and sometimes three-axle rigid chassis. In 1956 they were acquired by rivals Seagrave of Columbus, Ohio but continued to make their own independent designs. In 1979 Maxim was acquired by another fire engine maker, Ward La France.

*c.*1960 **Maxim fire engine.** Although a 1960s design, this conventional model was listed by Maxim right through the 1970s, with different grille pattern. Open cabs were rare by the 1970s, though acceptable in warm climates such as Florida where this unit operated at one time, although it served originally in Jenkintown, Pa, Fire Department. It has a Hercules 6-cylinder gasoline engine.

1973 Maxim Marauder pumper. Maxim did not introduce cab-over designs until 1959, but they have become an increasingly important part of the range. Engines were Cummins or Detroit diesels, and this truck has a 1600gpm Hale pump.

Mercedes-Benz

1979 Mercedes-Benz L-1113. Although of German design and appearance, this is an American-built truck as Mercedes began assembly in Jacksonville, Florida, in May 1979, using parts from their plant in Sao Paulo, Brazil. Tyres, electrical parts and bodies are entirely American made. The L series illustrated is the only US-assembled Mercedes-Benz, but the range runs from 25,000 to 32,000lb GVW. All use Mercedes-Benz diesel engines. A new plant at Newport News, Virginia, was scheduled to be opened in 1980.

Michigan

*c.*1972 **Michigan mobile crane,** made by a firm with a long history in construction machinery, though they did not make trucks until about 1963. This one is gasoline-powered, though most used diesel engines, usually Detroits. The company has long been owned by the Clark Corporation, and in the later 1970s the cranes began to carry the Clark name rather than Michigan.

New Holland

*c.*1976 **New Holland self-propelled baler.** Best known for its farm machinery, New Holland began in the town of that name in Pennsylvania, but has recently moved to New Holstein, Wisconsin. This vehicle can load and roll hay into a roll-shaped unit that is then pulled onto the back of the truck. It can also handle containers and crates. Power comes from Ford gasoline or diesel engines mounted to the right of the cab. New Holland is one of the few American trucks (Hagie is another) to use Hydrostatic drive.

Mobil

1975 Mobil street sweeper,a product of the American Hoist Company of St Paul, Minnesota, though since 1975 they have been owned by the Athley Company of Raleigh, North Carolina. A good many parts, including the engine, come from International, but Mobil make their own bodies and chassis.

Noland

1979 Noland electric truck. This Edgewater, Florida, company acquired the patterns, dies and parts of the defunct Pargo company early in 1979, and proceeded to market some of the Pargo truck and bus models under their own name. All are electric and come in four- and six-wheeled versions. This Model S-1 is rated to carry one ton.

J. & E. Olson

J. & E. Olson of Garden City, New York, started making aluminum bodies in 1947 and still specialize in this business today. They were acquired by the Grumman Aircraft Company in the early 1950s, and their vans sometimes bear the Grumman name in addition to, or instead of, the Olson name. Occasionally they carry the name of the chassis maker, but Olson is considered a make in its own right, as they do the final assembly and are therefore responsible to the US Government for seeing that safety requirements are met. Sizes range from half a ton to 19,000lb GVW. During the 1970s chassis have been mostly Ford, Chevrolet, GMC and International, with a few Mercedes-Benz.

1970 Olson van on Ford P-2000 chassis, with GVW rating of 7,000lb.

1978 Olson van on Chevrolet chassis,with GVW rating of 14,000lb.

Oshkosh

Oshkosh started in business in 1917 as the Wisconsin Duplex, a four-wheel-drive truck. It was in fact started by William Besserdich who had once been a partner in the FWD company, another Wisconsin company. (Oshkosh takes its name from the town of Oshkosh, Wisconsin, itself named from Indian Chief Oshkosh of the Menominee tribe.) Despite some incredibly bad years such as 1936 when they built but a single truck, Oshkosh has survived where many other truck makers have gone under, and today builds about a thousand units per year. Foreign assembly plants are in South Africa and Australia. There are many different lines of Oshkosh trucks, including fire engines for street and airport use, front-discharging concrete mixers, and cross-country tractors for military and civilian use. In the late 1970s they also made bus chassis.

1974 Oshkosh Model P 4 × 4 snowplow with Snogo equipment made by the Klauer Manufacturing Co of Dubuque, Iowa. Oshkosh snowplows are widely used in northern states of the US for street and airfield clearance.

1972 Oshkosh Model E 6×4 tractor with low-loading trailer. Available in two- and three-axle form, the Model E was no longer made for the US Market after 1974 as it was not redesigned to meet new brake regulations. It was still sold abroad, however, and is one of the models assembled in South Africa. This one has a Detroit diesel engine, though Cummins power was available as well.

1976 Oshkosh F-2365 6×6 US Army tractor. In September 1976 Oshkosh were awarded an army contract to build 1,192 of these heavy tractors, which are also known by their army designation of M-911. Intended mainly to haul tanks, they can also be used for transporting aircraft on, or off, highways. Engine is a 325hp Caterpillar diesel, similar to the largest option in the Model J. Note the auxiliary non-driven axle ahead of the driving tandem, which can be raised when the vehicle is lightly loaded. The standard semi-trailer used with this model has four axles.

1974 Oshkosh Model J oilfield truck, with giant tyres for operation on soft sand. This model was powered by Caterpillar diesel engines ranging from 250 to 325hp. Gross train weight with trailer was up to 180,000lb.

1977 Oshkosh P-15 8×8 airport crash truck fire engine. This massive machine was the largest fire engine in the world when introduced in 1977, and has two Caterpillar engines totalling 860hp. Fully loaded, it can reach a speed of 50mph from standstill in 50 seconds. First made for the US Air Force, it is now used by a number of municipal airports, including Los Angeles, which has two of them.

Ottawa

1977 Ottawa Commando 30. Ottawa was perhaps the first company in America (or anywhere) to make yard jockeys which are used to load and unload trailers from railroad trains in the so-called Piggy-back service which became increasingly important in the 1950s. It can be operated by one man as it has a hydraulically operated fifth wheel which can lift up and set down a trailer of up to 70,000lb GVW. Engines in early Ottawas were Fords, but are now mainly Detroit or Cummins diesels. Ottawas have been made in Ottawa, Kansas since 1959.

Pargo

1975 Pargo electric truck, on a similar chassis to that used for a golf cart, but it can haul a substantial 1,000lb load. Pargo made a wide variety of small electric vehicles including hotel buses and an unusual '19th hole car' which was a mobile bar to complement their golf carts!

In 1978 the Charlotte, North Carolina company went bankrupt and some of their assets were acquired by the Noland Car Co of Edgewater, Florida. (See page 41)

Peterbilt

Peterbilt was founded in 1939 when T. A. Peterman, a Washington State lumberman, decided that the logging industry needed special trucks and purchased the former Fageol truck plant in Oakland, California. Production barely got going before war broke out, but after the conflict Peterbilt became an important West Coast builder of heavy trucks, both on-highway and off-highway models for logging etc. Their first cab-over came in 1949, and in 1958 they were acquired by Pacific Car & Foundry, now Paccar, who already owned Kenworth. This has led to some similarity between the two makes, especially with the Kenworth L-700/Peterbilt 310 cab-overs. Peterbilt now have plants in Newark (California) and Madison (Tennessee) and build between 8,000 and 10,000 trucks per year.

1978 Peterbilt Model 310 6 x 4. This model is made in Kenworth's Quebec plant (which formerly made Sicard trucks) alongside the Kenworth L-700, which is the same truck apart from small differences in trim. The only model is the three-axle tandem drive illustrated, but engines can be Cummins or Detroit diesels from 230 to 350hp. This truck is carrying a vintage American La France fire engine.

1978 Peterbilt Model 359 tractor trailer unit. This is the top model of the range of Peterbilt conventionals and comes in a vast range of options which fill a 60-page catalogue. Engines are the usual diesels, up to 475hp, and transmissions are from Spicer, Fuller or Allison. Note the enormous radiator, over 1,400 square inches in area. Next to the 359 is its cab-over equivalent, the 352.

1971 Peterbilt Model 281 tractor trailer unit, similar to the Kenworth K-100. This is an earlier example of the Cab-over 'Pete' with front bumper narrower than more recent models. The steel spoked wheels were less common at any time than the usual aluminum ones.

1978 Peterbilt Model 352 tractor trailer unit. Like its conventional counterpart, the 352 cab-over comes with a vast variety of options and factory-offered colour schemes which are particularly popular with the owner driver. The two-axle version of this is known as the 282.

Pierce

1975 Pierce fire engine. The Pierce Manufacturing Company of Appleton, Wisconsin do not make chassis as such, but they re-work other makes to such an extent as to warrant consideration as a make in their own right. This unit is based on a Ford C-800 but has larger wheels and a Pierce-built body with six-man crew cab and Hale pump. Other brands on which Pierce have put their name include International, Dodge, FWD, Hendrickson, Oshkosh and Duplex.

Pirsch

1972 Pirsch fire engine. Another Wisconsin fire engine maker is Peter Pirsch of Racine, but unlike Pierce they do build chassis of their own as well as building equipment on other firms' chassis. As well as the cab-over illustrated, Pirsch make conventionals and 6-wheeled cab-overs, the latter usually fitted with ladder towers. This one has a 250hp Detroit diesel engine, but other makes can be specified by the customer.

Pitman

1970 Pitman ditch digger. This was built to compete with the better-known Gradall marketed by Warner & Swasey, and is mounted on a very short Dodge chassis and engine, though sheet metal is entirely Pitman's design and manufacture. Ray O. Pitman, who designed the Pitman Arm, hydraulic boom, now heads a new firm in Kansas named R-O, and the Pitman Arm is marketed by A. B. Chance of Sedalia, Missouri.

Rite-Way

1971 Rite-Way concrete mixer. This company dates back to about 1959 when American Rite-Way of Dallas, Texas, introduced the industry's first front discharge concrete mixer. This had the great advantage that the driver could put the truck exactly where the load was needed, instead of being involved in difficult reversing. By the mid-1970s the Rite-Way patents had run out, and at least eight other makers had adopted the front discharge system. Rite-Way used a rear-mounted Detroit or Cummins engine and a one-man cab mounted in the centre, not offset as are most one man cabs. This unit has 8 x 6 drive, but there were also 6 x 6, 8 x 4 or 6 x 4 models. Top speed was a surprising 70mph, and in the mid-seventies the price was $73,000. However this had exceeded six figures by the end of 1979.

Rohr

Rohr was a company with a diversity of interests which bought out Flxible-Southern, makers of Flxette buses and Flxible-Southern delivery vans in the late 1960s. They continued these under the former names, but in 1973 they renamed the vans Rohr, which title applied to about 1,500 vans built for UPS (United Parcel Service) during the 1973/74 season. The name then reverted to Flxible-Southern until 1976 when the division was sold to Leisure Time Products of Middlebury and Nappanee, Indiana who made a few more under the name LTP-Southern. More recent UPS vans of generally similar appearance have been made either by UPS themselves or by Airstream. (See page 7)

1974 Rohr van with 26,000lb GVW rating. This uses a Ford engine and chassis converted to forward control by Rohr.

Seagrave

Seagrave is one of the oldest names in American fire apparatus, dating from the 1880s. Motorized units first appeared in 1907. The Seagrave name and design were sold in 1963 to FWD and production transferred from their old home of Columbus (Ohio) to Clintonville (Wisconsin). Conventionals were discontinued in 1970, and their range now consists of two- and three-axle cab-forward models.

1960 Seagrave Model P fire engine, similar in appearance to the units being made during most of the 1970s. This one uses the Pierce-Arrow designed V-12 gasoline engine, but 1970s models had Detroit or Cummins diesel engines.

1979 Seagrave Model WB fire engine. This is the new low-profile Seagrave for the late '70s and '80s, powered by a Detroit diesel engine. The cab is at least three feet lower than on previous models. Engine and pump are mounted amidships. Seagrave also make a three-axle ladder truck and mount fire-fighting bodies on mass-produced chassis.

Spartan

1977 Spartan HH-1000 coal truck with options of Detroit or Cummins engines and a load capacity of 40 tons. GVWs are up to 140,000lb, depending on power and tyre size.

Spartan is one of the newest names in American trucking, being formed in 1975 by ex- Diamond-Reo suppliers and employees who tried to keep in business with a new line of trucks. In 1976 they built fifty fire truck chassis which they sold to FMC and John Bean to build bodies on, and later they made a small series of heavy coal trucks for open cast mining operations in Virginia. They also built the prototype for the revised Diamond-Reo that was subsequently put into production by Osterlund of Harrisburg. (See page 18.)

Sutphen

1978 Sutphen 6 x 4 ladder truck. This firm in Amlin, Ohio, began as wholesalers of fire equipment in 1890, and later were Pirsch dealers. They did not make their first complete fire engine until 1967, but have built up a reputation for high quality machines in the few years since then. While they make some two-axle jobs, their best-known product is this three-axle tower ladder and combination fire pumper. Its price in 1978 was about **$**250,000.

Terex

Terex was the division of General Motors that produced heavy dump trucks, scrapers and road construction machinery. Dating from 1972, in US Terex Vehicles were made at Hudson, Ohio, with the really large models coming from the GM diesel locomotive plant at London, Ontario.

1975 Terex TS-14 articulated scraper unit, with Detroit diesel 237hp V-8 engine and Allison automatic transmission. Load capacity is about 18 tons and it can be used as a bottom dump truck as well as a scraper. Its 11ft width does not bar it from American roads.

1974 Terex Titan 350-ton dump truck, the largest truck in the world. With an all-up weight of 600 tons, this monster is a record breaker in every direction. Power comes from a 3300hp GM Electromotive diesel engine normally used in railroad locomotives; capacity is 169 litres, and the weight of the engine alone is 16 tons. Transmission is by an alternator and four electric motors in each of the rear wheels. The tyres weigh more than $3\frac{1}{2}$ tons each and tower above the cab of a Kenworth conventional truck. The price was around $2\frac{1}{2}$ million dollars in 1974, but to build a replica today would cost 4 million. In fact, only one Titan was made, and is operating in an open-cast coal mine in British Columbia, supported by a fleet of fifty smaller dump trucks.

1978 Terex 50-ton dump truck. This is in fact one of the smaller Terex models, having a 525hp turbocharged V-12 Detroit diesel engine. It is 13 feet tall with a 13ft wheelbase, though overall length is over 27 feet. Empty weight is 71,000lb.

Tulsa

*c.*1975 **Tulsa 6×6 truck.** Tulsa Trucks Inc, named after the Oklahoma city where they are made, are conversions of Fords with a tandem rear axle and suspension, and also a powered front axle, giving six-wheel drive. Thus a truck rated at 23,000lb GVW is rebuilt to carry more than twice that amount. This one is based on a Ford F-600, and is used for removing and/or repositioning palm trees.

Universal Fire Apparatus

Universal Fire Apparatus is one of the many smaller builders of fire engines scattered about the United States who build custom equipment on other chassis, but the resulting product carries their own name rather than that of the chassis maker. Under their present name, Universal dates back to the early 1950s, but their history can be traced back to 1879 when they traded as Oberchain & Boyer. Today they make everything in the way of equipment, from nozzles to ladders, hoses, sirens, lamps and even the badges worn on firemen's uniforms.

1979 Universal 1000gpm Hale Pumper on a Hendrickson model 1871 chassis, supplied to Universal's home city of Logansport, Indiana.

1979 Universal 1250gpm Hale Pumper on a Spartan chassis, with Cincinnati cab and 1,000-gallon water tank of its own. This is the largest unit that Universal has made to date.

Wabco

Wabco is the current name of the former Westinghouse Air Brake Company which originally made air brakes for railroad trucks in the 19th century. The firm's Haulpak models were originally known under the name LeTourneau-Westinghouse, being designed by Robert LeTourneau who was responsible for many pioneering vehicles, mainly in the off-road category. His firms were split up after his death; the special vehicle division is known as Marathon LeTourneau and is still located at Longview, Texas, while the dump truck line went to Westinghouse, re-named Wabco in the 1960s. Capacities range from 35 tons to a monster 235 tons, and power outputs from 420hp to 2475hp.

1978 Wabco road scraper. Like their equivalents from Terex, these vehicles can also work as bottom dump trucks, first scraping off the earth and then leading it into their bodies for short hauls. They are licenced to run on the roads of many States, but only when empty. Power unit of this one is a Detroit diesel of 250hp.

1977 Wabco Haulpak Model 35 and Model 3200B dump trucks. These are respectively the smallest and largest trucks in the 1977 Wabco range. The Model 35 uses a 420hp Cummins or Detroit diesel engine, while the larger truck has a 2475hp General Motors EMD (Electro-Motive Drive) V-16. As with the largest Terex, transmission is through electric motors, though in the Wabcos case the motors are two in number.

Walter

1975 Walter road maintenance truck with the firm's patent Four Point Positive Drive. These smaller Walters had a choice of Ford gasoline, or Detroit diesel engines. The heavy brackets at the front are for mounting a snowplow in winter.

Walter is one of the oldest names still in business in America. The company was founded in Staten Island, New York by a Swiss immigrant, William Walter, and originally built chocolate-making machinery. His first automobile came in 1898, and truck production began in 1909. The factory was now in New York City (West 66th Street) and among the employees were two fellow Swiss, the Chevrolet brothers. The best-known early Walters were based on the French Latil 4-wheel-drive truck, and 4-wheel-drives have been among the most important Walter products up to the present. More recently, airport fire engines have been made, some with twin engines. Walters are now made at Voorheesville, in northern New York State.

1976 Walter Type B all-wheel-drive airport fire truck, with a 1,500gpm pump and 800-gallon tank. This is in service at the airport of Kinston, North Carolina.

Ward La France

Ward La France dates back to 1918 when they built a heavy truck chassis at their Elmira, New York, plant. The first fire engines came in 1931, and gradually this part of the business became predominant, though regular commercial trucks were made until about 1960. In the 1970s Ward La France built fire equipment on other chassis as well as their own custom pumpers, and in 1979 they acquired Maxim of Middleboro, Massachusetts. Despite the similarity of name and being made in the same town there has never been a company connection with American La France.

1976 Ward La France Ambassador 76 pumper, with 1,000gpm Hale pump. These units are powered either by Waukesha gasoline, or Cummins or Detroit diesel engines.

Wayne

*c.*1975 **Wayne sweeper.** A familiar sight in many cities, these sweeper units look like three-wheelers, but in fact have two rear wheels set close together. They are powered by gasoline engines, either Dodge or International. The factory is in San Jose, California, and the company is now part of the FMC conglomerate, some of the later units bearing the FMC brand name, or else FMC and Wayne combined.

Western Star

1975 Western Star conventional tractor trailer unit. This brand originated as a Western-orientated White product, built in Canada at Kelowna, British Columbia, and intended to compete with Freightliner. Unlike Freightliner, however, Western Star started as a conventional, and only added a cab-over in 1978. They use much aluminum and lightweight components, and offer highly decorated trim, with over 1,000 different paint schemes. Prices are high, starting at $73,000 in 1979 and going well into six figures. Detroit diesel or Cummins power is standard, with some very high power models of over 500hp being offered. About 1,000 Western Stars were built per year in the 1970s.

1975 White Road Expeditor 6 x 4 refuse truck. The Expeditor line was White's offering for the short-haul city market, especially for delivery work and the sanitation industry. They came in two- or three-axle form, with GVWs from 36,000 to 56,000lb. This one is fitted with a refuse collection body by Pak-Mor, a San Antonio, Texas, firm who were the originators of the hydraulically compacted body, designed to haul more before unloading.

White

White is a time-honored name in American trucking dating back to 1901 when they built their first steam delivery truck. Steamers were the sole product until 1909 when a line of gasoline trucks based on the French Berliet appeared. Since that time a great variety of vehicles including light trucks, fire engines and buses have been made. They were made in Cleveland, Ohio, right up to 1978 when the old plant became just a parts source and assembly at New River Valley, Virginia and Ogden, Utah. Diesel engines were first offered in 1934, and became standard in the early 1970s, though up to 1971 White offered 6-cylinder gasoline engines of their own make and also Chevrolet V-8 units. Autocar was acquired by White in 1953 and from 1976 the former White Construcktor construction industry conventional trucks were marketed as Autocar Construcktors.

1972 White Construcktor 6 x 4 concrete mixer, the model that was later made under the Autocar label. 6 x 6 chassis were made as well as 6 x 4s, and engines were usually Cummins or Detroit diesels.

1974 White Road boss tractor. This was the long version (122-inch BBC) of a unit that could be had also in two shorter sizes. Cummins or Detroit diesels were standard power units.

1977 White Road Boss II tractor, which succeeded the older type in 1975. Like its predecessors it came in three sizes, of which this is the shortest with 92-inch BBC measurement. A wide variety of accessories was offered, such as aluminum or chrome polished bumpers, aluminum alloy wheels, air-conditioning, etc. Power was usually Detroit or Cummins diesel, though one Caterpillar V-8 was offered in the late 1970s. Also tried experimentally, though never offered to the public, was the V-10 Mercedes-Benz diesel, used in about 15 trucks.

1977 White Road Commander II tractor trailer unit, the cab-over version of the Road Boss II. Single axle or tandem drive models were made, with Cummins or Detroit diesel engines up to 450hp, and GCVs up to 127,000lb.

Winnebago

***c.*1978 Winnebago delivery truck.** Winnebago of Forest City, Iowa, are the largest builders of the integral-type motor home in the United States, but in the later 1970s they have turned more and more to commercial truck and bus versions of their motor home shell, to take up the slack in motor home sales. They are available on both the Dodge and Chevrolet motor home chassis, both substantially modified, with V-8 gasoline engines, either Dodge, Chevrolet or International. This van is completely air-conditioned. Starting in 1979, Winnebago has been fitting Stirling-type engines to their AC unit generator sets, burning a variety of fuels, even paraffin.

*c.*1975 **Wollard aircraft service truck.** Located in Miami, Florida, Wollard is a specialist maker who devote all their production to aircraft service vehicles. These include scissors-lift trucks for delivery of foodstuffs or freight direct to a high airliner, or baggage-handling trucks such as this one, powered by GMC engine. Wollard use mass-produced chassis, but remodel the entire unit with their own sheet metal, altered steering, etc.

OTHER BRANDS OF AMERICAN TRUCKS

In addition to the many makers and marketing firms of trucks pictured in this book, dozens of other firms in the United States also marketed trucks or commercial-type vehicles in the seventies. The list is not complete, though it gives details of most other brands.

ABM, Springfield, Missouri, made custom chassis for motor homes, trucks, buses.

AG-CHEM, Jackson, Minnesota, made large three- and five-wheel fertilizer spreader trucks that could run off or on highway.

Allegheny, Huntington, West Virginia, made fire apparatus, some custom units with own brand name on them, though they did not make any trucks themselves.

Allegro, brand name of Tiffin Motor Homes of Red Bay, Alabama, also made a few commercial truck models using same motor home shell on Dodge base.

Alsea, Sioux City, Iowa, made chassis and complete units for other firms, notably the IHC delivery 1300/1500 chassis, chassis for various firms building motor homes and buses, and a few light trucks for themselves.

American Fire Apparatus, Battle Creek, Michigan, market fire apparatus, much of it their own design on chassis made for them.

American Snowblast make rotary powered snow blowers and plows in Denver, Colorado. Both modify and make their own chassis as well, some rear engined.

Arrow, Denver, Colorado, made a yard jockey type truck.

Austin-Western, Aurora, Illinois made crane carriers (and cranes).

Auto Specialties, St Joseph, Michigan, make small truck cranes.

AVCO, Tulsa, Oklahoma, and later in Nashville, Tennessee, make motor homes, some of which are made with blank type shell into trucks.

Badger, Winona, Minnesota, make cranes and crane carrier trucks, also known as BURRO brand, which is their rail type units.

Baker,Cleveland, Ohio, make small electric trucks, oldest electric firm (since 1899) still in business in US.

Banner, Elkhart, Indiana, a motor home builder also build some truck models using same body shell.

Barth, Milford, Indiana, another motor home builder make trucks using their motor home shell, some of them mobile hospital type vehicles with three axles.

Battronic, Boyertown, Pennsylvania, make electric trucks, delivery type.

Beall, Grand Rapids, Michigan, make aircraft tractors for pulling baggage, and other airport vehicles (electric).

Bean, Tipton, Indiana. John Bean make fire apparatus, some with own badge on them though they buy all their chassis.

Big Joe, Lincolnwood, Illinois, make small trucks for use inside plants.

Boyertown Body, Boyertown, Pennsylvania, make truck bodies, but many also have their integral bodies, and only the Boyertown badge on them.

Broderson, Lenexa, Kansas, make small truck cranes on own chassis.

Bros make road scrapers and other machinery; some of them borderline trucks.

Bruco, Brumbaugh Body's brand name, of Altoona, Pennsylvania, make fire apparatus and mobile hospitals, classrooms and rescue vehicles. Buy chassis, but own brand badge on many of vehicles.

Bucyrus-Erie, South Milwaukee, Wisconsin, make cranes, on truck chassis, marked with their badge name. Most are bought elsewhere.

Buick, Flint, Michigan, market a long chassis model primarily for use in professional service (funeral) cars.

B & Z, Long Beach, California, later of Signal Hill, California, make small electric vehicles and trucks.

Cadillac, Detroit, Michigan, long have made a commercial car chassis primarily used by firms who make professional car bodies, but other custom body firms also mount pickup bodies on them as well.

Calavar, Sante Fe Springs, California, make a custom fire tower, usually on a FWD chassis, though not always, but with their own trim and badge.

Capacity, Longview, Texas, is a yard jockey type truck.

Cardox, variously listed as in Chicago or Philadelphia, made fire-fighting equipment and marketed a few complete trucks with their own badge on them.

Cargo King, is made by Cochrane Equipment Co, Salinas, California. This truck is designed to haul freight to and from airplanes and to load and unload it.

Carpenter, Mitchell, Indiana. This bus body-builder has also built a few truck models using same body shell, Use Carpenter badge, but firm builds no chassis of their own.

Case, Racine, Wisconsin, market the Daimler-Benz and made UNIMOG in America with Case name painted on.

Chair Yacht, Shoshoni, Wyoming, made a small gas-powered vehicle designed to carry either a person confined to a wheelchair, in a vehicle he could drive from the chair, or also used as small flat bed truck, with other driver.

Champion, Champion Carriers of Dallas, Texas, made yard jockey type truck, crane carriers and other specialized trucks.

Champion, Made by American Coleman Co of Littleton, Colorado, took over production of yard jockey model only for Dallas firm in late 70s.

Champion, Bremen, Indiana, and Dryden, Michigan, make motor homes and, using same chassis and shell, make buses and trucks also, all on basic Dodge chassis.

Champion, Darley & Co of Melrose Park, Illinois, market Champion brand fire apparatus, some custom units with own nameplate and design, chassis made for them.

Clark of Buchanan, Michigan, make a variety of specialized trucks, mostly crane carriers. Some have dual badges such as Michigan, Austin-Western and Lima, all divisions of firm, and each vehicle made in their specific plant.

Clarke, Muskegon, Michigan, make small gas-powered trucks and sweepers.

Cline, Kansas City, Missouri, make huge mining trucks, also a yard jockey type unit, and other specialized vehicles. Products sold under ISCO name, 1972–78.

Club Car, Augusta, Georgia, make small electric vehicles.

Coachman, Middlebury, Indiana, make motor homes on Dodge chassis. Sometimes, they are made as trucks with same body shell.

Commaco, Kansas City, Kansas, made small truck cranes.

Compass Industries of Hermosa Beach, California made small electric trucks.

Condec, made by Consolidated Diesel-Electric of Old Greenwich, Connecticut, made specialized trucks for use at airports, fire fighting, military use, etc.

Condor, Sante Fe Springs, California, is a truck crane, made for utility use by Calavar firm, using a FWD chassis with different trim, nameplate.

Cortez, formerly the **Clark Cortez,** it is a medium-sized motor home on its own chassis, and was made at Battle Creek, Michigan early in 70s, later in Kent, Ohio. Some models made as trucks using same body shell.

Crown, Los Angeles, California, make fire trucks (including own chassis), motor homes, large van type trucks and buses. Also make bodies only.

Cruiser, Minneapolis, made a truck-mounted street sweeper, using modified IHC chassis with some of own sheet metal and rebuilt frame.

Curtis, Oakland, California, made fire apparatus, some custom units with own badge on them, though chassis purchased.

Deere, John, East Moline, Illinois and Dubuque, Waterloo, Iowa, make large articulated type trucks used in conjunction with scrapers, dump and other bodies.

Difco, also known as **Differential,** Findlay, Ohio, make large mining trucks.

Drott, Wausau, Wisconsin, make a large truck crane, and also smaller rough terrain cranes. Owned by Case, but separate division.

Dumptor, made by Koehring Co of Milwaukee, is a large dump type mining truck.

Duralite, Baltimore, Maryland, are truck body makers who also built some delivery vans of medium size of own design on modified mass-produced chassis.

Eagle. Small electric vehicles made by Legend Golf Car of Dallas, Texas.

Easton Car, made in Easton, Pennsylvania, mainly self-powered dump vehicles used inside factories, but make bodies as well.

Ecolotec, Clark, New Jersey, market a line of sanitation trucks with own bodies, on modified chassis of other makers, but with own brand name on them, and revised sheet metal.

Electric Pony Express, Tulsa, Oklahoma, make small electric trucks.

Electric Vehicles, South San Francisco, California, market modified vehicles with with electric battery propulsion.

Electromotion, Bedford, Massachusetts, marketed mass-produced small trucks with electric battery propulsion.

Electro-Van, Austin, Texas, market van type vehicles of other firms with their own electric propulsion system, and badge on vehicle. Firm name, Jet Industries.

Elgin of Elgin, Illinois, are one of the oldest brands of motorized truck-mounted sweepers, dating back to 1905. They normally use International gasoline engines. Elgin also market a complete sanitation line including Packmaster bodies made by former Leach company of Milwaukee, and sweepers on International Cargostar chassis, but with Elgin name on front of cab.

Elwell-Parker, Cleveland, Ohio, long a supplier of electric type light trucks.

Emergency One, Ocala, Florida, make fire apparatus and armoured cars, rescue trucks, some with own badge on them, though they only modify and do not make chassis.

Erickson, Minneapolis, Minnesota, make a yard jockey and small industrial use trucks.

Eriez, Erie, Pennsylvania, make magnetic cranes and also sweepers, on modified mass produced chassis, sometimes with their own badge on them.

Etnyre, Oregon, Illinois, make unusual vehicles such as asphalt spreaders, sprinkler trucks and the like, with own design and chassis.

Euclid, Euclid, Ohio, make large mining trucks, scrapers.

E-Z-GO, Augusta, Georgia, made small electric and gasoline-powered trucks.

Fabco, Hayward, California, make specialized custom trucks, such as yard jockeys, utility trucks, fire trucks, trucks to wash tunnels, and special conversions of trucks into all-wheel-drive, or even add chain-drive if that is what is wanted.

Fairmont Railway Motor of Fairmont, Minnesota, are better known for their railroad maintenance of way vehicles, and also convert mass-produced cars and trucks for use on rails or roads. They make small trucks to serve airports too, such as tow trucks, platform trucks and the like.

Farrar of Woodville, Massachusetts, make fire apparatus, some trucks with their own name on them, though they buy all the chassis.

Fire Trucks of Mount Clemens, Michigan make what their name implies, complete fire trucks, including their own chassis.

Frank of Pampa, Texas, market huge drilling rigs on truck bases. It is likely that CCC make the truck chassis for them, but all have FRANK nameplate.

Fyr Fox of Houston, Texas, Market fire apparatus, some custom units with their brand name on them, though they buy all their truck chassis.

Galion of Galion, Ohio, make all sorts of construction vehicles including a medium-sized truck crane.

Geiman of Marion, Ohio, make truck cranes.

General Safety Equipment of North Branch, Minnesota, make fire apparatus, some custom units with their brand name on them, though they buy all their truck chassis.

Gerlinger, Dallas, Oregon, is owned by Caterpillar and make truck crane carriers and the cranes on them as well, and their own chassis.

Getman of South Haven, Michigan, make all types of truck to work **inside** underground mines. These are mining dump trucks, cranes, personnel carriers.

Goodbary Engineering Co of Tulsa, Oklahoma, make huge mining dump trucks.

Grumman market a wide variety of truck bodies and also trucks with their name on them and truck chassis much modified by them. Some of their products have dual names along with that of their divisions, such as HOWE, OLSON or OREN. They make trucks, buses, fire trucks, ambulances and delivery vans.

Hanson, a Tiffin, Ohio firm, market truck cranes.

Harley-Davidson, of Milwaukee, Wisconsin and York, Pennsylvania, the well-known motorcycle maker, also make three- and four-wheel small trucks, both gas and electric.

Harsco, made by Bowen and McLaughlin of York, Pennsylvania, make military vehicles.

Heil Co of Milwaukee, Wisconsin, market a small sanitation truck fitted with their own body, but actually a truck built by Hagie.' It is sold by Heil with their nameplates on it. It is rear engine, gasoline-powered.

Henderson, Manchester, Iowa, market a large 20-ton three-wheel truck that has a fertilizer spreader body on it.

Hesston of Hesston, Kansas, are better known for their farm machinery, some of it self-propelled, but make a small gas-powered truck as well.

HMK Marketeer of Redlands, California, was brief name of firm that succeeded the Westinghouse Marketeer brand. Made electric trucks and buses.

Holmes, The Ernest Holmes Co of Chattanooga, Tennessee, are the largest makers of wrecker or breakdown cranes in the world, and they also market a large wrecker-like crane that can be used by railroads to re-rail cars or rolling stock, some of which are fitted to roll over rails or highway. While basically CCC truck bases, the **Holmes** name and guarantee go with each unit, and are the only nameplate shown on vehicle itself.

Hough of Libertyville, Illinois, is now owned by International Harvester, and now source of their **Payline** construction type vehicles, though Hough name still appeared into 70s. Among trucks made are articulated type scrapers and dump trucks.

Howe of Anderson, Indiana and Martinez, California, are vendors of fire apparatus, some of custom units with their nameplates only, though they make no trucks of their own.

Huber of Marion, Ohio, are primarily tractor and construction machinery makers, but they do make a truck-like crane unit.

Hyster, whose main offices are in Portland, Oregon, but whose plants are located coast to coast, market a number of truck-like vehicles, including a 9-wheel roller all on rubber tires that is fitted with express body, and can carry tools to work as well as do a job there. They also make straddle trucks as used in lumber yards, which carry load beneath truck, and tow trucks, light truck cranes and the like.

Imperial of Rancocas, New Jersey, sold fire apparatus, mostly with own name on vehicle though usually a chassis built for them.

Isco, Kansas City, Missouri, built large mining dump trucks, combination rail-highway trucks and yard jockeys and specialized tow trucks. Before 1972 , and after 1978, known as CLINE.

Jaco is brand name for Jack Cocke Co of Mobile, Alabama, who market fire apparatus, some of it own name badge, though they buy all chassis.

Jacobsen of Racine, Wisconsin, make small gas-powered trucks.

Kalamazoo of Michigan town of same name, made fire crash trucks and other specialized municipal or rail service vehicles.

Kayot of Indianola, Iowa, make boats and motor homes, some of latter delivered as trucks with motor home bodyshell. Dodge chassis is underneath, but their name is on unit.

K-D, Waco, Texas, make rough terrain trucks, small electric and gas trucks.

Klein make articulated fire crash trucks for airports. Their name is on truck, but they purchase, not build it: only build body.

Knickerbocker, Jackson, Michigan, make small tow and industrial type trucks.

Koehring, Milwaukee, Wisconsin, name appears on truck cranes and huge Dumptor trucks.

Laher, Oakland, California, make small electric trucks.

Landau, Anaheim, California, a builder of motor homes also build a truck model using same basic body shell. Chassis are modified quite heavily but purchased.

Lectra Haul, Tulsa, Oklahoma, make huge dump trucks for mining and construction. They have electric drive, hence name, though powered by diesels.

Lektro, Warrenton, Oregon, make small electric trucks.

Lima, perhaps best known for their steam locomotives in years gone by, now make among other things, at their Lima, Ohio, plant, large truck-mounted cranes.

Link Belt, Cedar Rapids, Iowa, make truck cranes. Some of truck bases, perhaps all, are purchased, but only their brand name appears.

Linehauler, San Jose, California, make a yard jockey type truck.

Lockheed, The well-known California aircraft maker also marketed an unusual many wheel off highway truck called the Dragon Twister, that could operate over very rough surfaces, off roads.

LTP-Southern, Evergreen, Alabama, made a few UPS van type trucks, as well as truck bodies and buses.

Lyncoach, Troy, Alabama, made a door to door delivery truck with their nameplate but chassis was purchased. It also went under name of **Lyn Arrow.**

Madsen, of Bath, New York and later Allentown, Pennsylvania, made a yard jockey and other specialized trucks, also buses.

Manitowoc, of Manitowoc, Wisconsin, make large truck-mounted cranes. At one time they sold their product under **Bay City** nameplate.

Marathon-LeTourneau make unusual vehicles for all sorts of mining, construction and ecology purposes. Some are truck-like cranes or scrapers. They are located in Longview, Texas.

Marion Body Works of Marion, Wisconsin, make all sorts of truck bodies, among them fire apparatus, some of these only with **Marion** nameplate. They do purchase all truck chassis though.

Mars of Minneapolis, Minnesota, make small electric trucks and sweepers.

Maxon, Los Angeles, California, are truck body makers, and recently started making their own truck chassis, sanitation trucks as well, similar to competing Master with many fiberglass parts.

Medical Coaches, Oneonta, New York, make large van type integral unit trucks used for mobile hospitals, rescue trucks, offices and schools.

Melex, Boca Raton, Florida, assemble knock-down kits of small electric vehicles including trucks, the kits made in Poland, but using American-made batteries and tires.

Micky, High Point, North Carolina, make a moving van type truck of medium to heavy size. They also make truck bodies. Chassis are purchased.

Midas, Elkhart, Indiana, better known for their mufflers, also make a delivery type van.

Milford, Milford, Connecticut, make large crane carriers and truck cranes.

Mobile Electric of Waterloo, Iowa, make small electric trucks.

Motor Coach Industries of Pembina, North Dakota and Winnipeg, Canada, make mostly intercity buses and are owned largely by the Greyhound Corporation, However the bus body shell is sometimes used for travelling classroom, display showroom, etc work, with the windows blanked out.

Murty Brothers, Portland, Oregon, make large truck cranes.

National Water Trucks, of Midway, Washington, are highly unusual specialty truck of large size, fitted with half-cab and made primarily complete with body for use in water service during construction. It is partly built with old military truck parts, some reconditioned.

Newell Coach of Miami, Oklahoma, built motor coaches, buses and sometimes trucks using same shells, as used on others, both front and rear engine models.

Nordco-Marketeer, of Redlands, California, is most recent name of this maker of small electric vehicles including trucks.

Northwest, Green Bay, Wisconsin, make large and medium-sized truck cranes.

Nu Star, Minneapolis, Minnesota, made by Erikson, is a brand of yard jockey type truck, also small electric trucks.

Oldsmobile, Lansing, Michigan, made their front-wheel-drive three-axle chassis for use as a motor home, bus or truck, gas-powered.

Ome, Ocala, Florida, made a small gasoline-powered truck.

Onan Westcoaster, Stockton, California, made small three-wheel gas-powered van type trucks, Also plant in Minneapolis, Minnesota.

Orbie, Washington, DC, market a line of sanitation pickup trucks, from small three-wheel vehicles on up to medium-sized units. Most chassis were probably built for firm, though their name is what appears on the units.

Oren, Vinton, Virginia, make fire apparatus. Many have their brand name on truck portion though this is usually built for them. They once built trucks though.

Ortruk, St Louis, Missouri, made huge dump trucks for mining and construction.

Otis made small gasoline and also electric trucks, some of them, the Westcoaster models made in Stockton, California, formerly the Onan brand, and some in Cleveland, Ohio.

Paramount of Riverside, California, make water trucks, partly of surplus parts and reconditioned parts from old military trucks. All are all-wheel-drive.

Pem-Fab, Rancocos, New Jersey, build fire apparatus, including parts of truck chassis.

Penn Versatile Van of Chicago, Illinois, built delivery type integral vans. Usually chassis was purchased, but their badge is on vehicle.

Pettibone, also know as Pettibone-Mercury, and Pettibone-Miliken, a Chicago-based firm market large truck cranes as well as smaller rough terrain models, also Mercury brand small tow trucks. Some chassis are purchased.

Phillips, Los Angeles, California, made small gas trucks used in inter-plant service.

Pickwick, Cedar Rapids, Iowa, made custom chassis for motor homes, buses and trucks of van type.

Pierce-Pacific, Portland, Oregon make large truck-mounted cranes and truck carrier also, as well as large logging trucks.

Plymouth of Plymouth, Ohio, is no relation to the Chrysler built Plymouth, but is a large off highway mining type dump truck.

Plymouth, Detroit, Michigan. Certain Dodge light trucks and minibuses are sold under the brand name of this other Chrysler Corporation passenger car.

Prevost is really a Canadian company from Sainte Claire, Quebec, but they have a US preparation plant at Madison, New Jersey, where vehicles are altered for US market. Like Motor Coach Industries, they are essentially bus makers, but without side windows, these vehicles can be used as mobile classrooms, etc.

Quality Fire Apparatus, Eastaboca, Alabama, make fire apparatus, some units of custom design with their nameplate, though they buy chassis.

Ray-Go, Minneapolis, Minnesota, make many types of construction machinery, among them truck-like articulated scraper units: also used as dump trucks.

Ray-Go-Wagner, Portland, Oregon, are custom builders of mining machinery and trucks for use in mining and construction.

Raymond of Greene, New York, and also Santa Clara, California, make small electric trucks used inside plants or dock warehouses.

Revcon of Fountain Valley, California, make motor homes and sometimes use their motor home shell in a truck unit. Both front and rear engine designs were made.

Rex was marketed by the Rexnord Co of Milwaukee, Wisconsin. They were thinly disguised Diamond-Reos complete with **Rexnord** built bodies, and **Rex** nameplates and different grills, sold through **Rexnord** organization.

Rickel of Salina, Kansas, make mostly quite large vehicles, big all-wheel-drive dump trucks for use off highway, and the **Terra Gator,** which is a large three-wheel truck fitted with fertilizer spreader body and mounted on high floatation tires.

R-O Products of Olathe, Kansas, make mobile cranes designed by R.O. Pitman who was formerly with Pitman, and also all-wheel-drive conversions of regular trucks.

RPC a product of Midland-Ross of Roxboro, North Carolina, is a brand of straddle-type trucks carrying the load beneath the vehicle rather than on it. They are borderline trucks in usual sense.

Ryder, Jacksonville, Florida, the truck rental company, had ten of the so called Pacemaker type design tilt cab truck made for them by Hendrickson to design of an Oregon man, Dean Hobgenseifken. No more were made after the initial ten in 1973–75.

Sanford, Syracuse, New York, dates back many years in commercial truck field, but in 1970s only made fire apparatus, some of it small three-wheel gas models, much on backs of mass-produced brands, but some custom units with Sanford name, though these too were purchased chassis.

Scat, is brand name of the City Fire Equipment Co, of Perrysburg, Ohio. They sometimes also badge their equipment CFE. As might be imagined they make fire apparatus, mostly bodies on back of mass-produced trucks, but they also make small four-wheel gas fire trucks of Cushman size, all fitted with Scat brand badge.

Silent Hoist, Brooklyn, New York, make a small truck crane.

Smithco, Wayne, Pennsylvania, make quite small trucks, both three- and four-wheel, all gasoline powered.

Star, Tacoma, Washington, make small trucks, gas and electric.

Strick Cab Under was an unusual truck consisting of a chassis and cab which could be slotted under a large trailer to give maximum payload within a given length. Power was from a Cummins V-8 diesel located behind the 4-ft high cab. However, it was widely criticized by drivers on grounds of safety, and opposed by the unions. The Strick Corporation of Fairless Hills, Pennsylvania, did not proceed with the design, introduced in 1978.

Taskmaster, Dundee, Illinois, make fire apparatus, and also water tank trucks. Some are custom designs fitted with their own brand name and trim, though they have all the chassis built for them.

Taylor-Dunn, Anaheim, California, make small electric trucks.

Thomas, High Point, North Carolina, mainly build buses and bus bodies, but also make a few units using bus shell, and fitting them out as trucks.

Tiffin of Tiffin, Ohio, make truck cranes.

Towers, Freeburg, Illinois, make fire apparatus, including crash trucks, Some vehicles have their badge and appearance, though they buy all chassis.

Towmotor make small industrial trucks, also make tractor-type tow trucks, that can and do pull trailers.

Trackmobile, made by Whiting Corp. of Harvey, Illinois, make a combined locomotive and tractor-type truck. It has both wheels to run on rails and rubber-tired wheels to run on roads, and it can haul several railroad cars, or also push vehicles or haul semi-trailers as well.

Travco of Brown City, Michigan, through 1978 and later, made vehicles in Nachadoches, Texas. They make motor homes, some with blank sides for use as trucks; others are used as blood banks, mobile hospitals, sales rooms, offices and the like.

Tryco Mfg Co Inc of Decatur, Illinois make three- and four-wheeled sludge shooters for spreading fertilizers on fields, and other agricultural applications. Engines are Ford V-8 gasoline, Detroit or Cummins diesels, and the four-wheelers have four-wheel-drive and steering. The largest Super T Four has a payload of 18 tons, and even the three-wheeler has a load capacity of 7 tons.

Unimaso, Gardena, California, are largest makers of small trucks whose work is to paint lines down the lanes of public highways. Their machines literally travel millions of miles per year in this work and also roll to and from their work over public highways as well. Gasoline engines are used.

Unit Crane of New Berlin, Wisconsin, make medium-sized truck-mounted cranes, and quite often the truck carrier is also fitted with UNIT badge, though in fact most are supplied by FWD.

Vanguard, sometimes called **Sebring Vanguard**, or also **Citivan,** made a small electric van called the **Citivan.** They were located in Sebring, Florida.

Van Pelt, of Oakdale, California, make fire apparatus, some of the custom designs with their own label on them, though all chassis are made for them.

Vintage Reproductions of Oakland Park, Florida, make replicas of cars and trucks of the early 1900s, including early Ford and Olds models, some of which are fitted with a pickup back deck.

Ward, Conway, Arkansas, are primarily bus and bus-body makers, but like most such firms they also make some of their bus shells into trucks instead, with blanked out side windows.

Western States Fire Apparatus of Cornelius, Oregon, make fire apparatus bodies, and also market some custom design trucks with their badge alone on them. They are bought, however. except for body work.

Westinghouse, Marketeer, Redlands, California, were small electric trucks and buses.

Wittenburg, Midway, Washington, made trucks largely out of rebuilt, or surplus unused parts of old military trucks, combined with new fiberglass parts, into new all-wheel-drive trucks of various sizes, gas and diesel powered.

Work Horse, Riverside, California, make small gasoline trucks and industrial trucks.

Yale, of Willow Grove, Pennsylvania, and also Forest City, Arkansas, made mostly lift trucks, but some of their larger models, could well be described as truck cranes mounted on a truck chassis and are of the rough terrain type.

Young of Lancaster, New York, are makers of fire apparatus, and also make their own chassis for their custom units as well, assembled from various well-known parts, plus their own individual sheet metal and fiberglass.

Y-W stands for Yankee-Walter, of Los Angeles, California, who made medium-sized crash trucks for use at airports, some of which were custom designs with their own nameplate, though in fact all chassis were made for them.

ACKNOWLEDGEMENTS

The majority of the photographs are from the author's collection, but he is indebted to the following who have kindly lent him additional photographs:

American Coleman Co, Crane Carrier Corp, Major Fred Crimson (US Army, Retired), Custombilt Trucks Inc, Fiat-Allis, G. N. Georgano, General Motors, Ferdinand Hediger, F.W.D. Corporation, Hendrickson Mfg Co, Ibex Inc, Jeep Corp, Tony D. Kelly, Kenworth Truck Co, Mack Trucks Inc, Marmon Motor Co, Marmon Transmotive, Master Truck, Nolan Co, Noland Car Co Inc, Oshkosh Truck Corp, Peter Pirsch & Sons Co, Seagrave Fire Apparatus Division (F.W.D. Corp), Universal Fire Apparatus Corp, Wabco Construction & Mining Equipment Group, Walter Motor Truck Co, Ward La France Truck Corp and R. A. Wawrzyniak.

INDEX